Percy Faraday Frankland

**Agricultural Chemical Analysis**

Percy Faraday Frankland

**Agricultural Chemical Analysis**

ISBN/EAN: 9783337060978

Printed in Europe, USA, Canada, Australia, Japan

Cover: Foto ©berggeist007 / pixelio.de

More available books at **www.hansebooks.com**

# AGRICULTURAL

# CHEMICAL ANALYSIS

BY

## PERCY F. FRANKLAND, Ph.D., B.Sc., F.C.S., F.I.C.

ASSOCIATE OF THE ROYAL SCHOOL OF MINES,
PROFESSOR OF CHEMISTRY IN UNIVERSITY COLLEGE, DUNDEE,
FORMERLY LECTURER AND SENIOR DEMONSTRATOR OF CHEMISTRY IN THE
NORMAL SCHOOL OF SCIENCE AND ROYAL SCHOOL OF MINES,
SOUTH KENSINGTON MUSEUM

FOUNDED UPON

"Leitfaden für die Agricultur-Chemische Analyse"

VON DR. F. KROCKER.

## London
MACMILLAN AND CO.
AND NEW YORK
1889

*First Edition 1883*
*Reprinted 1889*

*Printed by* R. & R. CLARK, *Edinburgh.*

# PREFACE.

AT the outset I intended that these pages should be merely a translation of Dr. F. Krocker's excellent handbook, entitled *Leitfaden für die Agricultur-Chemische Analyse*, which, through his generosity, was placed at my disposal. Soon, however, I came to the conclusion that a somewhat more extensive treatment of the strictly agricultural portion would be desirable in a work of the kind published in this country.

I have therefore described more exhaustively the examination of substances connected with agriculture ; whilst, to avoid much increasing the size of the volume, I have almost entirely passed over the General Principles of Analytical Chemistry, with which I have assumed the reader to be already conversant.

Besides the invaluable aid derived from the above work of Prof. Krocker's, I have obtained much useful assistance from the French handbook of Agricultural Chemical Analysis, by Prof. Grandeau ; whilst in the section on Water Analysis I am indebted to my father for the loan of the woodcuts.

It is hoped that this book will be of assistance to analytical chemists and in laboratory teaching, inasmuch as I have attempted to satisfy a want which I have felt myself in conducting classes in Agricultural Chemistry at the Normal School of Science.

GROVE HOUSE,
PEMBRIDGE SQUARE, W.

# CONTENTS.

## I. QUALITATIVE ANALYSIS.

### DETECTION OF MINERAL BASES AND ACIDS.

## II. QUANTITATIVE ANALYSIS.

### I. BASES.

#### GRAVIMETRIC AND VOLUMETRIC DETERMINATION.

## II. ACIDS.

## III. DETERMINATION OF FREE ACID OR ALKALI IN SOLUTION.

## IV. ELEMENTARY ORGANIC ANALYSIS.

## V. EXAMINATION OF SOILS.

## VI. ANALYSIS OF PLANTS AND VEGETABLE STRUCTURES.

## VII. ANALYSIS OF MANURES.

## VIII. ANALYSIS OF MILK AND OTHER DAIRY PRODUCE.

## IX. ANALYSIS OF WATER.

# APPENDIX.

# AGRICULTURAL CHEMICAL ANALYSIS.

## I. QUALITATIVE ANALYSIS.

### Detection of the Bases and Acids of Common Occurrence in Agricultural Products.

## SECTION I.

#### SUBSTANCES SOLUBLE IN WATER.

### A. Examination for Bases.

1. TEST the reaction of the solution with litmus-paper.
    - *a.* The liquid is **acid** :[1] presence of a free acid or of an acid salt, or of a salt containing a **weak base** (*e.g.* alumina and iron).
    - *b.* The liquid is alkaline : presence of a free **alkali** or of a salt with **strong base and weak acid** (*e.g.* alkaline **carbonates**, silicates, or sulphides).
    - *c.* The liquid is **neutral.**

[1] In this case it is necessary to examine for compounds of phosphoric acid, mentioned in Section II., as these, on treatment with acids, are partially dissolved by water and precipitated by ammonia (thus, calcic phosphate which has been treated with sulphuric acid). A small portion of the solution is therefore tested for phosphoric acid with ammonic molybdate, and if phosphoric acid is present, the further examination is proceeded with as in Section II. p. 8.

S    3 <    B

2. Treat a portion of the original concentrated solution or some of the solid substance, moistened with water, with caustic potash. The presence of **ammonia** is then ascertained by the smell, turmeric paper, etc.

3. To one half of the solution, whether neutral, acid, or alkaline, now add hydrochloric acid.

 a. An **evolution of gas** takes place : indication of the presence of **carbonates, sulphides** (sulphites, acetates).

 b. A gelatinous **precipitate** indicates silicic acid ; even if no precipitation takes place, small quantities of silica may yet be present in the solution ; and, if suspected, the solution should be evaporated to dryness, ignited and taken up with dilute hydrochloric acid ; the silica will then be left as an insoluble residue.

4. Add **ammonia** to a portion of the acidulated solution, from which the silica has been separated as above. If no precipitate is formed the ammoniacal-solution is further examined as under (5) ; on the other hand, if precipitation take place, the filtrate is treated according to (5), whilst the precipitate ($Fe_2O_3$, FeO, $Al_2O_3$) is examined as follows :—

 a. The precipitate is white, gelatinous, and soluble in caustic soda .    .    .    . Alumina.

 b. The precipitate is brown, gelatinous, and insoluble in caustic soda .    .    . **Ferric oxide.**

 c. The precipitate is of a greenish colour .    .    . } **Ferrous** oxide.

Irrespectively of the colour of the precipitate, treat a small portion of the original solution with potassic ferricyanide ($K_6Fe_2Cy_{12}$) ; if, by the precipitation of Turnbull's blue, the presence of ferrous oxide be thus established, the solution must be boiled with a few drops of strong nitric acid to convert the FeO into $Fe_2O_3$, then add ammonia in excess, and filter.

A portion of this precipitate is to be heated with dilute sodic hydrate, filtered, and the filtrate acidulated with HCl, then ammonia added until the alkaline reaction is restored ; a white gelatinous precipitate is formed . **Alumina.**

5. The filtrate obtained from the whole precipitate, or, in the absence of any precipitation by ammonic hydrate, the ammoniacal-solution, is treated with **ammonic oxalate,** a white precipitate . . . **Lime.**

The precipitation is rendered complete, the solution warmed, filtered, and to the filtrate some more ammonic oxalate is added to make sure that the lime has been completely separated.

6. To a small portion of the filtrate some **hydric disodic phosphate** is now added, a white crystalline precipitate . . . . . **Magnesia.**

*Note.*—If the precipitate be flocculent, it should be dissolved in a few drops of dilute HCl, the solution gently heated and rendered alkaline with ammonia, and then the sides of the glass vessel containing it should be scraped with a glass rod ; if magnesia be present, the precipitate will settle on cooling where the glass has been scratched.

> *a.* **Magnesia absent.**—If, on evaporating a few drops of the filtrate (to which no sodic phosphate was added), a fixed residue be obtained, the whole of the filtrate from the calcic oxalate is evaporated to dryness and ignited until no more ammoniacal fumes are evolved.

α. No fixed residue remains. } **Absence of Alkalies.**

β. A fixed residue remains. Dissolve in the smallest possible quantity of water acidulated with hydrochloric acid, and divide the concentrated solution into two parts.

The larger portion is tested with platinic chloride or tartaric acid for . . . Potassium.

The smaller portion is tested on platinum-wire in the Bunsen-flame for . . . . Sodium.

b. **Magnesia present.**—The solution, after separating the lime, is evaporated to dryness, the ammonia-salts are driven off by ignition, and the residue is then taken up with water. Heat to boiling, and then add a solution of baric hydrate until alkaline, heat and filter.

The filtrate contains the alkalies and the excess of baric hydrate.

Warm and add ammonic carbonate until the excess of baric hydrate is completely precipitated; filter off the baric carbonate and evaporate the filtrate (which now contains only the alkalies and the excess of ammonic carbonate) to dryness, ignite until the ammonia-salts are volatilised, take up with a little dilute hydrochloric acid, and examine the solution for alkalies as described above, a.

*Note.*—As the double salts of platinic chloride, with the chlorides of calcium and magnesium and sodium, are soluble, it is possible, by using sufficient platinic chloride, and if only chlorides be present, to detect the potassium without previously separating the lime and magnesia.

## B. Examination for Acids.

1. A portion of the substance is treated with **hydrochloric acid**; an evolution of gas may be due to— **carbonic acid, sulphuretted** hydrogen (also sulphurous anhydride and acetic acid); the sulphuretted hydrogen is recognisable by its smell and action on lead-paper.

2. To a portion of the original solution add **baric chloride** and then dilute hydrochloric acid :—

   a. A precipitate insoluble in HCl is formed  .  .  . } **Sulphuric Acid.**

   b. The white precipitate dissolves in HCl with effervescence  .  .  . **Carbonic Acid.**

   c. The white precipitate dissolves in HCl without effervescence, and is reformed on addition of ammonia  .  .  . **Phosphoric Acid.**

3. The original acidulated solution is tested with **ammonia and magnesia mixture,** or with ammonic molybdate for  .  .  . **Phosphoric Acid.**

4. A fresh portion of the original solution is treated with argentic nitrate. A white precipitate, insoluble in nitric acid, readily soluble in ammonia :

**Hydrochloric Acid.**

5. A portion of the original substance, or some of the concentrated aqueous solution, is treated with strong sulphuric acid, and the mixture allowed to cool ; then a solution of **ferrous sulphate** is carefully added, a **brown ring** is formed at the surface of contact  .  **Nitric Acid.**

6. A portion of the original solution is evaporated to dryness with HCl and ignited, the residue is insoluble in HCl  .  .  .  .  . **Silicic Acid.**

# SECTION II.

SUBSTANCES INSOLUBLE IN WATER, BUT SOLUBLE IN ACIDS.

IN the analysis of agricultural products the following must be principally taken into account :—

  *a.* Carbonates of lime and magnesia.

  *b.* Phosphates of lime, magnesia, iron, and alumina.

  *c.* Oxides of iron, aluminium, and manganese.

  *d.* Silicious compounds decomposable by acids (slags, cements, silicious ashes, etc.), decomposable sulphur compounds.

  *e.* Compounds difficultly soluble in water (gypsum, lime).

  *f.* Organic compounds, carbon, etc., which are destroyed on ignition.

Treat a portion of the substance with HCl, avoiding unnecessary excess, boil for some time, dilute with water and filter if the solution is not clear.

  *a.* Evolution of gas in-} Carbonic Acid,
     dicates    .    .  } Sulphuretted Hydrogen.

  *b.* The solution is more or less yel-} Ferric Oxide.
     low coloured .    .    . }

     (Test especially with potassic ferro, and ferricyanides.)

  *c.* Gelatinous precipitation .    . Silicic Acid.

Test the solution in the first place for ferrous iron with potassic ferricyanide.

If ferrous iron be present in the HCl-solution, it

must be oxidised with a few drops of strong nitric acid before the separation of the other bases can be proceeded with.

If the substance contains silica, the latter must be separated by evaporating the acid-solution to dryness and igniting. If this precaution be neglected, it must be borne in mind that silica may be found in the ammonia precipitate and also with the alkalies.

1. The hydrochloric-acid-solution (to which nitric acid may have been added for the oxidation of the iron) is now warmed, and a slight excess of ammonia added. The solution should contain a due proportion of ammonic chloride.

   *a.* There is no precipitate.

The solution is further tested with **ammonic oxalate** for . . . . . **Lime.**

After heating and filtering off the latter, **hydric disodic phosphate** precipitates . . **Magnesia.**

   *b.* A precipitate is formed. Heat gently for some time until the smell of ammonia is no longer perceptible, then filter.

  I. The filtrate is examined as under (*a*) for **lime** and magnesia.

II. Examination of the ammonia precipitate—

  1. Dissolve a portion of the precipitate in HCl.

    *a.* Treat a portion of this solution with sodic hydrate in excess, boil for some time, filter, acidulate the filtrate with HCl and then add ammonia in slight excess. A white gelatinous precipitate: **Alumina.**

    *b.* Test a second portion of the solution with potassic ferrocyanide for . . . **Ferric Oxide.**

2. Fuse a portion of the precipitate with sodic carbonate on platinum-foil, a **bluish-green mass** is obtained

**Manganese.**

3. Dissolve a small portion of the precipitate in nitric acid, add excess of ammonic-molybdate-solution, and warm gently . A **yellow precipitate**: Phosphoric Acid.

If phosphoric acid is present, the precipitate may con tain, besides the bodies already mentioned, also the **phosphates of the alkaline earths.**

### Phosphoric Acid is present.

The further examination is varied according as the precipitate contains only traces of ferric oxide and larger quantities of the phosphates of the alkaline earths (readily recognisable by the white or yellowish-white colour), or only traces of phosphoric acid and principally ferric oxide (apparent from the brown colour of the precipitate), thus :—

*A.* The precipitate is **light coloured** and contains **much phosphoric acid** (bone-ash, guano, vegetable ashes). There are present—$Fe_2O_3$, $P_2O_5$, CaO, MgO.

Dissolve the precipitate in HCl, add ammonia to the *cold* solution until the liquid is just alkaline, and then acetic acid in excess. **Ferric and Aluminic phosphates** are **precipitated.** In solution there remain :—CaO, MgO, and possibly $P_2O_5$. Filter off and wash.

### 1. Examination of the Filtrate.

*a.* Add to the solution, which is acid with acetic acid, **ammonic oxalate.** The **lime** is precipitated as calcic oxalate $\left( \begin{cases} CO \\ \quad CaO_2 \\ CO \end{cases} \right)$ ; warm and filter.

*b.* The filtrate, which may contain MgO and $P_2O_5$, should now be rendered alkaline with ammonia. The **magnesia** is precipitated as **ammonic magnesic phosphate**, which is filtered off and washed with water containing ammonia.

*c.* The filtrate may now still contain $P_2O_5$. Treat with **magnesia mixture**. The phosphoric acid is precipitated as **ammonic magnesic phosphate**.

## 2. Examination of the Precipitate.

*a.* Boil with sodic-hydrate-solution and filter. The ferric phosphate is decomposed, leaving ferric hydrate as a residue; the aluminic phosphate is dissolved unchanged. Test the residue by dissolving in HCl and adding potassic ferrocyanide, a precipitate of Prussian blue  .  **Iron** (as phosphate).

*b.* Acidulate the filtrate with HCl, and then add ammonia in slight excess, a white gelatinous precipitate  .  .  **Aluminium** (as phosphate).

*B.* The precipitate **is brown, and contains** $Fe_2O_3$, $Al_2O_3$, **MnO, together with traces of phosphoric acid.**

For the examination of soils and similar substances, in which, after elimination of the silica, the chief problem is the separation of iron, aluminium, phosphoric acid, manganese, together with the alkaline earths and alkalies, the following method offers advantages, especially in the more perfect separation of the manganese and of the alkaline earths :—

Divide the original acid solution into two parts, A and B :

*A.* After diluting largely, nearly neutralise the free HCl with sodic carbonate, heat to boiling, and then add excess of a saturated solution of sodic acetate ; by this means the free HCl is replaced by free acetic acid :—

> *a.* By boiling the whole of the $Fe_2O_3$, $Al_2O_3$ and $P_2O_5$ are precipitated.
>
> *b.* To the filtrate add at a temperature of about 60° C. a few drops of bromine, and allow to stand twelve hours in a warm place, a brown precipitate of hydrated peroxide of manganese $\begin{cases} MnO\,(OH) \\ MnO\,(OH) \end{cases}$ Manganese.
>
> *c.* The filtrate from (*b*) precipitate with **ammonic oxalate** . . . . . **Lime.**
>
> *d.* The filtrate from (*c*) precipitate with hydric disodic phosphate . . . **Magnesia.**

*B.* In this portion examine for the alkalies (see p. 4, *b*). By separating the precipitate obtained with ammonia, and dissolving this in nitric acid, the phosphoric acid can be tested for with ammonic molybdate.

The acids are detected in the ordinary course of analysis indicated above; thus, $CO_2$, $SH_2$, $P_2O_5$, $SiO_2$. $SO_2(OH)_2$ should be tested for in the HCl-solution with $BaCl_2$.

# SECTION III.

## SUBSTANCES INSOLUBLE IN WATER AND ACIDS.

THE substances which are most commonly met with, and have to be examined under this head, are the silicates not decomposed by HCl (clay, sand, amorphous silica), the difficultly-soluble or insoluble sulphates of the alkaline earths, organic matters, carbon, etc. The latter are easily recognisable by their external appearance and combustibility, whilst the amorphous silica is soluble in sodic carbonate, from which, by acidulation with HCl, and then evaporating to dryness and igniting, the $SiO_2$ may be separated and identified.

For the closer examination of such a residue insoluble in HCl, one of the following methods may be employed according to the nature of the substance :—

1. The residue contains **clay** (silicate of alumina)— **treatment with $SO_2$ $(HO)_2$** should be resorted to. The finely-powdered substance is mixed in a platinum dish with one part of water, three parts of strong $SO_2$ $(HO)_2$ being then added ; the mixture is then heated until the greater part of the acid has been volatilised. After repeating this operation several times, the residue is boiled with HCl, filtered, and the solution examined according to Section II. above.

The residue is **sand and amorphous silica.** The latter is soluble in caustic and carbonated alkalies, whilst the sand is left undissolved by these reagents.

2. **Treatment with sodic carbonate solution.**— The residue insoluble in acids should be subjected to prolonged boiling with a concentrated solution of sodic carbonate.

    *a.* **Sulphates of the** alkaline **earths** (excepting baric sulphate) are hereby converted into soluble alkaline sulphates and insoluble carbonates of the alkaline earths. The sulphuric acid is to be looked for in the solution, the alkaline earths in the residue.

    *b.* Amorphous **silica**, which may be either due to the decomposition of a silicate with $SO_2(HO)_2$, or present as such in the original substance, is dissolved. The filtrate is tested for $SiO_2$ by evaporating the acidulated solution to dryness, heating to 110° C.; then, on taking up with dilute HCl, there remains insoluble    .        .        . **Silica.**

3. **Decomposition by fusion—**

    *a.* With fusion-**mixture** $CO(NaO)_2$, $CO(KO)_2$.—This method is adopted in the case of silicates which are not to be examined for alkalies.

    The finely-powdered substance is mixed with about three times its weight of fusion-mixture in a platinum crucible, and the mixture covered over with a thin layer of the same flux. The crucible cover is now put on and the bottom heated gently until the mass is fused, then more strongly until the decomposition is complete. Allow to cool. In order that the fused mass may be removed from the crucible, the latter should be now heated until the margin of the cake begins to fuse, and then allowed to cool. Then sufficient water is added to cover the cake, and on warming gently the latter readily detaches itself and can be transferred to a dish. Soften the mass with water, add an excess of HCl, and separate the $SiO_2$ by evaporating in the manner already described. The filtrate is examined further according to Section II. above.

    *b.* With **baric carbonate** or fluorides; in the case of silicates which are to be tested for alkalies.

        *α.* With baric carbonate.—Mix the finely-powdered substance with five or six times its weight of baric carbonate, and heat strongly in a platinum crucible in the manner already described. The cindery mass obtained on cooling is to be heated with very dilute HCl until completely decomposed, and the $SiO_2$ separated by evaporation, etc. The baryta is then removed from the filtrate by means of $SO_2(HO)_2$, and the examination for bases then continued as described above.

β. With ammonic fluoride.—Warm the finely-powdered substance with six times its weight of $(NH_4)F$ and a little water in a platinum crucible, then heat gradually to dull redness until no more fumes are evolved. Treat the residue with $SO_2(HO)_2$ and volatilise the excess of the latter. Dissolve in HCl; any insoluble residue must be further treated with $(NH_4)F$.

γ. With aqueous hydrofluoric acid.—Cover the finely-powdered silicate with aqueous but tolerably concentrated HF, digest for some time on the water-bath, then treat with dilute $SO_2(HO)_2$, evaporate to dryness on the water-bath, and finally heat more strongly until the excess of $SO_2(HO)_2$ has been driven off. The cooled mass contains the bases as sulphates.

δ. With gaseous hydrofluoric acid.—Spread out 2 or 3 grms. of the finely-powdered mineral in a shallow platinum dish, and moisten with dilute $SO_2(HO)_2$; then place the dish in a leaden vessel, on the bottom of which there is an inch stratum of fluorspar made into a paste with concentrated $SO_2(HO)_2$; the mixture is then exposed to the action of the HF vapour at a temperature not exceeding 60° C. The residue contains fluo-silicates and sulphates. After addition of $SO_2(HO)_2$ in excess, and nearly volatilising the whole of the latter, the residue is to be treated with HCl and water. The solution contains the bases.

# II.  QUANTITATIVE  ANALYSIS.

The Quantitative estimation and Separation of the
Compounds most frequently met with in Agricultural Products.

THE proportions in which different bodies occur are determined either by weight (gravimetric analysis) or by
measure (volumetric analysis).

*A.* In gravimetric analysis the separation of the
bodies from their solutions takes place in definite forms
which permit of being weighed.  The separations or " pre-
cipitations " must be complete, and the precipitates possessed
of a definite composition, in order that the body in question
may be determined either directly or indirectly by calculation.

In complex substances, in which several bases and
several acids are present, it is necessary to take into
account the method of separation.

For the performance of these determinations certain
apparatus is required, which will be specified and described
in the course of the work.

*B.* Volumetric analysis.—Many bodies, both bases
and acids, can be determined volumetrically with equal
accuracy and with far greater rapidity than by weight.  In
this case the quantity of the reagent in solution, which is

required to separate or to convert the body in question into some definite compound, is measured by volume.

The completion of the operation must be recognisable by definite and sharp reactions, *e.g.* change of colour or the formation of a precipitate. (The Indicator.) The solutions employed as reagents are of definite strength and are known as titrated or standard solutions. They should be prepared in considerable quantity, so as to be always ready for use.

It is convenient to make standard solutions of what is known as normal strength. In order to simplify the calculations one equivalent in grms. of the body in question is made up to 1000 c.c. of solution. Each c.c. then contains ·001 equivalent. In the case of bodies possessed of very powerful reaction, solutions are used containing but $\frac{1}{10}$ equivalent in 1000 c.c. (Decinormal solutions.) In some cases it is convenient to use definite percentage-solutions; thus, to make 10 grms. of the substance up to 1000 c.c. of solution, then each c.c. contains ·01 grm. of the substance. (Empirical solutions.)

Solutions of indefinite strength are also employed, which require standardisation from time to time.

The operations used in volumetric analysis are based upon one of the following classes of reactions :—

1. Reactions of **saturation**. (*E.g.* Alkalimetry and Acidimetry.)

2. Reactions of precipitation.

3. Reactions of oxidation and reduction.

The most important apparatus necessary for volumetric determinations are :—

    *a.* Measuring flasks, of 1000, 500, 300, 200, 100, 50 c.c. capacity.

b. **Graduated cylinders**, of 500 c.c. capacity graduated into 5 c.c., and 100 c.c. graduated to 1 c.c.

c. **Pipettes**, delivering a definite volume of liquid, of 100, 50, 25, 10 c.c. capacity.

d. **Graduated** pipettes, one of 20 c.c. graduated in fifths, and one of 10 c.c. graduated in tenths.

e. **Burettes**, stoppered and with pinchcock delivery-tubes.

## I. BASES.

## I. ALKALIES.

A. **Gravimetric and Volumetric Determination.**

### 1. Potash.

Potash can be determined as—potassic sulphate, potassic nitrate, potassic chloride, and potassic platinic chloride.

a. **Potassic sulphate** and **potassic chloride** are ignited gently and weighed.

| | | | | | |
|---|---|---|---|---|---|
| $SO_3$ 80·00 | . | 45·92 | Cl 35·46 | . | 47·55 |
| $K_2O$ 94·22 | . | 54·08 | K 39·11 | . | 52·45 |
| | | 100·00 | | | 100·00 |

A mixture of potassic chloride and potassic sulphate (also the nitrate and carbonate) is converted into the acid potassic sulphate by evaporating to dryness with excess of $SO_2(HO)_2$. This readily loses half its sulphuric acid and its water of hydration, if the mass be repeatedly ignited with a few

c

pieces of pure ammonic carbonate in a crucible with the lid on. The neutral sulphate, of the above composition, then remains.

b. Potassic **nitrate** is dried at 120° C., ignited, and weighed.

| | | | |
|---|---|---|---|
| $N_2O_5$ | . . | 108·06 | . . 53·41 |
| $K_2O$ | . . | 94·22 | . . 46·59 |
| | | | 100·00 |

c. Potassic **platinic chloride.**—The chloride, free from all ammonia-salts, is precipitated with platinic chloride. If the potassium is also combined with other acids, these compounds are converted into the chloride, as described in the Qualitative Analysis, p. 4.

The concentrated aqueous solution is treated with platinic chloride (which must be wholly soluble in 80 % alcohol) in excess, and evaporated almost to dryness on the water-bath; the precipitate is then treated with alcohol (80 %), and, after several hours' digestion, thrown upon a weighed filter, and washed with alcohol. The precipitate is then dried at 100° C. until of constant weight. The precipitate is potassic platinic **chloride** $2KCl + PtCl_4$:

| | | | |
|---|---|---|---|
| 2KCl | . . | 2(74·57) | . . 30·72 |
| $PtCl_4$ | . . | 336·30 | . . 69·28 |
| | | | 100·00 |

N.B.—The platinum can be readily recovered from this precipitate and from the alcohol by boiling the residues together with the alcoholic washings in a large basin with strong caustic soda. The platinum separates as a black powder, which, after being filtered off and well washed, can be converted into platinic chloride.

## 2. Soda.

Soda can be weighed as sodic sulphate, also as sodic chloride, sodic nitrate, and sodic carbonate, provided only one of these latter salts be present. The weighing, together with the mode of converting into the sulphate, is carried out in the manner described for Potash:

| | | | | | | |
|---|---|---|---|---|---|---|
| $SO_3$ | . | 80 | . | 56·34 | $Cl$ 35·46 . 60·66 | |
| $Na_2O$ | . | 62 | . | 43·66 | $Na$ 23 . 39·34 $= 53\,Na_2O$ | |
| | | | | 100·00 | 100·00 | |
| $N_2O_5$ | . | 108 | . | 63·53 | $CO_2$ . 44 . 41·51 | |
| $Na_2O$ | | 62 | . | 36·47 | $Na_2O$ . 62 . 58·49 | |
| | | | | 100·00 | 100·00 | |

## Separation of Sodium and Potassium.

The mixture of the two alkalies is weighed as alkaline chlorides; the other alkaline salts, if present, having been previously converted into chlorides. The potassium is then determined as potassic platinic chloride; the quantity of potassic chloride is calculated from the platinum precipitate; after subtracting its weight from that of the mixed chlorides, the difference is sodic chloride.

Note.—In a mixture of the pure chlorides the alkalies can be determined *indirectly* by calculation, after determining the chlorine in a given weight of the mixed chlorides by means of standard argentic-nitrate-solution and potassic chromate (see p. 51); thus, let S = weight of mixed chlorides, C = the weight of total chlorine, then KCl = 4·63S − 7·63C; the NaCl is found by difference.

### 3. Ammonia.

Ammonia is best determined as—

a. **Ammonic chloride,** by evaporation of the solution on the water-bath in a tared glass or platinum dish, and subsequent drying at 100° C. until of constant weight. This temperature must not be exceeded, otherwise a loss of ammonic chloride would be occasioned. 100 parts contain 31·8NH$_3$, or 26·18N.

b. **Ammonic platinic chloride.**—In the presence of several acids, to the concentrated acid-solution, which must be free from potash, add platinic chloride, and proceed as described in the case of potassium.

The precipitate $(NH_4Cl)_2 + PtCl_4 = 7.67$ % $NH_3 = 6.32$ % N.

| | | | | |
|---|---|---|---|---|
| 2(NH$_3$) | . | . | 34 . | . | 7·67 |
| 2(HCl) | . | . | 72·92 | . | 16·45 |
| PtCl$_4$ | . | . | 336·30 | . | 75·88 |
| | | | 443·22 | | 100·00 |

The precipitate is collected on a weighed filter and dried at 100° C. until constant. On ignition chlorine and ammonic chloride are evolved, the platinum remaining as a porous mass (platinum-sponge), from the weight of which the ammonia may also be calculated—

$$100Pt = 17\cdot48NH_3 = 14\cdot40N.$$

c. By **volatilisation** of the ammonia from its salts.

(α). At a high **temperature** by distillation with sodic or calcic hydrate—

Introduce the substance in a tube $1\frac{1}{2}$ ins. long and $\frac{1}{4}$ in. diam. into a flask containing a moderately strong solution of sodic hydrate or milk of lime; place the flask in an inclined position on a wire-gauze, and connect it by means of a glass-tube with a small condenser. The delivery-tube of the latter is to be connected with a tubulated receiver, to the tubulus of which is attached a small U-tube.

A measured quantity of normal acid is now poured into the receiver, and a small portion of the same into the U-tube. The normal acid in the receiver should be tinted with a little litmus. As soon as the drops distilling over no longer momentarily colour the liquid blue, the operation is complete.

After carefully washing the normal acid in the receiver and U-tube into a small flask, the free acid is determined by means of normal alkali; the ammonia is calculated by difference; or, if the acid in the receiver was hydrochloric acid, the ammonia may be determined gravimetrically by means of platinic chloride as above.

The operation may also be conducted in the following manner with the use of more simple apparatus:—1 or 2 grms. of the ammonia-salt are introduced into a flask (A) of about $\frac{1}{2}$ lit. capacity, furnished with a doubly-bored cork. One boring of the latter serves for the reception of a

pipette containing caustic soda, and the outer extremity of
which is closed with a pinchcock ; into the second boring is
inserted a doubly-bent glass-tube leading to the bottom of
a bottle (B) of about 1 lit. capacity.   The latter is also
provided with a doubly-perforated cork, the one opening
of which receives the delivery-tube mentioned above,
whilst a U-tube is fitted into the other.

The latter is filled with glass beads moistened with
water, and on them rests a small piece of red litmus-paper.

In the bottle B is placed a measured quantity of normal
acid tinted red with litmus, over which the delivery-tube
opens.   As soon as the apparatus is air-tight, caustic potash
is allowed to enter A by opening the pinchcock ; the
contents are then boiled until the drops of liquid distilling
over no longer impart a transient blue colour to the
normal acid.   The apparatus is left for an hour to insure
the complete absorption of the $NH_3$ ; the normal acid is
then titrated with normal alkali and the $NH_3$ calculated
(see Alkalimetry).

(The whole of the free $NH_3$ present in a liquid passes
over with the first third of the liquid that is distilled.
The volatilisation of the $NH_3$ proceeds more readily if a
current of steam be passed through the liquid, as by this
means the inconvenient bumping of the liquid is prevented).

β. At the ordinary temperature by means of
lime or caustic potash.   (Schlössing's
Method.) (Applicable only in the case of ammoniacal solutions of 30 to 40 c.c. volume, and
containing not more than ·3 grms. $NH_3$.)—

A measured volume of standard acid is placed in a
shallow dish (about 5 ins. diam.), above which is supported

on a tripod-stand a small basin (about $2\frac{1}{2}$ ins. diam.) containing the solution of the ammonia-salt. The whole is then placed upon a plate containing sufficient mercury that a bell-jar or beaker can be made to close air-tight over it By raising the glass, or by admission through a tubular in the latter, the lime-water or caustic potash, as the case may be, is added to the ammoniacal solution, and the apparatus closed again as rapidly as possible. After standing forty-eight hours the acid in the dish is titrated with standard alkali, and the $NH_3$ calculated from the acid neutralised.

       *d*. With **small quantities** of ammonia-salt.—By decomposing the ammonia-salt WITH A STRONG SOLUTION OF BROMINATED SODIC HYPOCHLORITE (Eau de Javelle) and estimation of the $NH_3$ :—

          *α*. **By** titration, applicable in the absence of any other substances oxidisable by chlorine. The ammonia-salt is decomposed with a definite quantity of the brominated hypochlorite, standardised with arsenious acid, the undecomposed hypochlorite being afterwards determined by a second titration with arsenious acid. The brominated hypochlorite decomposed corresponds to a definite quantity of $NH_3$.

**Preparation of the** brominated hypochlorite.— Chlorine is passed to saturation through a solution containing 1 part $Na_2CO_3$ in 15 parts $OH_2$ kept cold with ice ; to the liquid so obtained strong NaHO is added until it is soapy to the touch, and 2 to 3 grms. of Br. per litre are further added each time before use.

**Standardisation of the brominated** hypochlorite.

—This is effected by means of a solution of sodic arsenite, containing in the litre 4·95 grms. of $As_2O_3$ (free from $As_2S_3$) $= \frac{1}{40}$ molecular weight in grms. in 1000 c.c. The following is the reaction which takes place with the hypochlorite :—

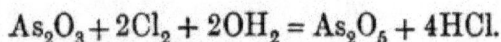

$$As_2O_3 + 2Cl_2 + 2OH_2 = As_2O_5 + 4HCl.$$

Dilute 10 c.c. of the brominated hypochlorite with water, and then add the $As_2O_3$-solution from a burette until a drop taken out upon a glass rod gives no blue colouration with iodised starch paper.

As from the above 1 molecule of $As_2O_3$ requires, for its conversion into $As_2O_5$, 4Cl, 4Br, or 2O, therefore $\frac{1}{40}$ molec. of $As_2O_3$ is equivalent to $\frac{1}{10}$ atom of Cl, and every c.c. of the above $As_2O_3$-solution used corresponds to $\frac{}{10.000}$ atomic weight of Cl in grms.

Moreover, since 3Cl are requisite for the decomposition of one molec. $NH_3$ and liberation of N $(3Cl + NH_3 = 3HCl + N)$, therefore 3 c.c. of the $As_2O_3$-solution are equivalent to $\frac{1}{10.000}$ molec. weight of $NH_3$ (·0017) in grms., and 1 c.c. to ·000566 grms. $NH_3$, or ·000466 grms. N.

**Preparation of the iodised** starch-paper.—This should always be freshly prepared by mixing 1 part of starch with a small quantity of cold water until a smooth paste is obtained; then about 150 or 200 times its weight of boiling water are added, and after being allowed to cool and settle, the clear liquor is poured off and mixed with a few drops of potassic iodide; strips of pure filter paper are soaked in the liquor, and the paper so prepared should be used in the damp state, as it is then most sensitive.

Dry iodised starch-paper can be prepared by boiling 3 grms. of starch with 250 c.c. of water; a solution of 1 grm. of potassic iodide and 1 grm. crystallised sodic carbonate are then added, and the whole diluted to 500 c.c. ; after subsidence the filter-paper is soaked in the clear liquid and dried in an air-bath at a low temperature.

**The titration.**—Add the brominated hypochlorite from a stoppered burette to the diluted solution (about 1 : 1000) of the ammonium-salt until the effervescence due to evolution of nitrogen gas ceases.   Allow to stand for ten minutes, then dilute with water free from ammonia, and run in the solution of sodic arsenite from a burette until the liquid ceases to give a reaction with the iodised starch-paper.

The volume of brominated hypochlorite used is equivalent to a definite volume of $As_2O_3$-solution, and if the quantity of the latter actually used be subtracted, then the difference will represent the volume of $As_2O_3$-solution equivalent to the ammonia decomposed.

### β. By the volume of nitrogen liberated (Knop's Azotometer)—

A weighed quantity of the ammoniacal compound is dissolved in a little water and introduced into Knop's azotometer (Fig. 1), which consists of a glass cylinder, B (20 to 25 c.c. capacity), which is fused to the bottom of an outer and larger glass flask-shaped vessel, A (150 to 200 c.c. capacity).   The ammoniacal solution is

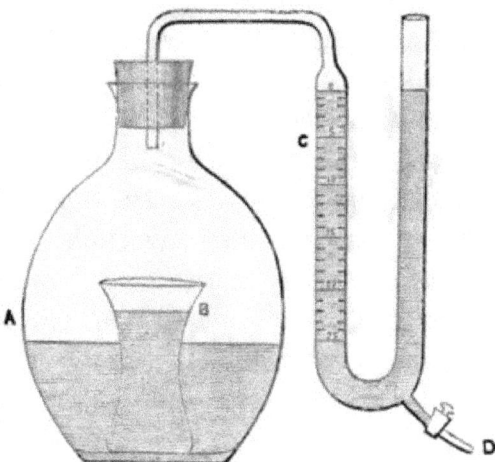

Fig. 1.

placed in the inner cylinder, whilst into the space between the inner and outer vessels 50 c.c. of the hypobromite solution (prepared by dissolving 100 grms. of sodic hydrate in 1250 grms. of water and adding 25 c.c. of bromine) are introduced, and the neck of the outer vessel is then closed with a perforated india-rubber-stopper. A glass-tube passes through the latter, and its U-shaped extremity is graduated into cubic centimetres. This graduated tube is filled with water to the zero mark, which is placed at the top, the level of the water being the same in both limbs. The flask-shaped vessel is now immersed in water, so that both it and its contents may acquire a uniform temperature. By judicious shaking, the ammoniacal liquid in the cylindrical vessel is thrown into the outer vessel, where it comes in contact with the hypobromite. Evolution of nitrogen now takes place, which is accompanied by a depression of the water in the graduated limb of the U-tube. When the reaction is finished, the vessel is again immersed in water to reacquire its former temperature, and the level in the two limbs of the U-tube is equalised by drawing off some of the water by means of a stopcock, D, placed at the bottom of the tube. The volume of nitrogen liberated is then read off on the graduated limb. Owing to the nitrogen being slightly soluble in the liquid from which it was evolved, a small correction has to be made. The extent of this correction is ascertained from the following table, constructed by Dietrich, in which the amount of liquid used is supposed to be 60 c.c. (50 c.c. hypobromite and 10 c.c. water.)

| c.c. | | | | | | | | | |
|---|---|---|---|---|---|---|---|---|---|
| 10 ·28 | 20 ·53 | 30 ·78 | 40 1·03 | 50 1·28 | 60 1·53 | 70 1·78 | 80 2·03 | 90 2·28 | 100 2·53 |
| 9 ·26 | 19 ·51 | 29 ·76 | 39 1·01 | 49 1·26 | 59 1·51 | 69 1·76 | 79 2·01 | 89 2·26 | 99 2·51 |
| 8 ·23 | 18 ·48 | 28 ·73 | 38 ·98 | 48 1·23 | 58 1·48 | 68 1·73 | 78 1·98 | 88 2·23 | 98 2·48 |
| 7 ·21 | 17 ·46 | 27 ·71 | 37 ·96 | 47 1·21 | 57 1·46 | 67 1·71 | 77 1·96 | 87 2·21 | 97 2·46 |
| 6 ·18 | 16 ·43 | 26 ·68 | 36 ·93 | 46 1·18 | 56 1·43 | 66 1·68 | 76 1·93 | 86 2·18 | 96 2·43 |
| 5 ·16 | 15 ·41 | 25 ·66 | 35 ·91 | 45 1·16 | 55 1·41 | 65 1·66 | 75 1·91 | 85 2·16 | 95 2·41 |
| 4 ·13 | 14 ·38 | 24 ·63 | 34 ·88 | 44 1·13 | 54 1·38 | 64 1·63 | 74 1·88 | 84 2·13 | 94 2·38 |
| 3 ·11 | 13 ·36 | 23 ·61 | 33 ·86 | 43 1·11 | 53 1·36 | 63 1·61 | 73 1·86 | 83 2·11 | 93 2·36 |
| 2 ·08 | 12 ·33 | 22 ·58 | 32 ·83 | 42 1·08 | 52 1·33 | 62 1·58 | 72 1·83 | 82 2·08 | 92 2·33 |
| 1 ·06 | 11 ·51 | 21 ·56 | 31 ·81 | 41 1·06 | 51 1·31 | 61 1·56 | 71 1·81 | 81 2·06 | 91 2·31 |
| Nitrogen evolved, „ absorbed | Nitrogen evolved, „ absorbed | Nitrogen evolved, „ absorbed | Nitrogen evolved, „ absorbed | Nitrogen evolved, „ absorbed | Nitrogen evolved, „ absorbed | Nitrogen evolved, „ absorbed | Nitrogen evolved, „ absorbed | Nitrogen evolved, absorbed | Nitrogen evolved, absorbed |

($\gamma$) Colorimetrically.—By means of Nessler's solution.   (See Water Analysis.)

**B.—Determination of the Alkalies alkalimetrically.**
(See Acidimetry and Alkalimetry.)

## C.—Determination of the Alkalies by means of the Specific Gravity of their Solutions.

In the case of *pure* solutions of potash, soda, or ammonia, the quantity of alkali contained can be ascertained by means of the specific gravity. (See tables in the Appendix.)

## II. ALKALINE EARTHS (Lime—Magnesia).

### 1. Lime (CaO).

### A. Gravimetric Determination.

Lime is weighed as—

|  | Calcic Carbonate |  |  | or Calcic sulphate. |  |  |
|---|---|---|---|---|---|---|
| $\{$ | $CaO$ | . . 56 | $CaO$ | . . 56 | . . | 41·18 |
| $\{$ | $CO_2$ | . . 44 | $SO_3$ | . . 80 | . . | 58·82 |
|  |  | 100 |  | 136 |  | 100·00 |

The precipitation is generally effected by means of ammonic oxalate $\left( \left\{ \begin{array}{l} CO(NH_4O) \\ CO(NH_4O) \end{array} \right) \right.$; as calcic oxalate $\left( \left\{ \begin{array}{l} CO \\ CO \end{array} CaO_2 \right) \right.$, which by gentle ignition is converted into calcic carbonate ($COCaO_2$).

1. If the calcium-salt is soluble in water, add to the hot solution an excess of ammonic oxalate, together with a little free ammonia; leave the precipitate to subside in a warm place for six hours, and then filter the clear liquid. Wash the precipitate three times by decantation, and then bring the whole upon the filter and wash repeatedly

there. The complete subsidence of the precipitate is much facilitated by employing the ammonic oxalate solution hot. Should the precipitate exhibit any tendency to pass through the filter, then no fresh liquid should be poured on to the filter until the latter is empty.

When the precipitate has been thoroughly washed on the filter, the latter is dried, the precipitate carefully detached and placed in a platinum crucible ; the filter-paper is burnt on platinum wire and then dropped into the crucible. The crucible should then be heated very carefully to incipient redness, allowed to cool in a dessicator, and then weighed. If the contents, on being moistened with water, impart a brown colour to turmeric paper, (showing that the calcic carbonate has been further converted into quicklime), a small piece of pure ammonic carbonate should be added, and the crucible again ignited very gently until the ammonia is all driven off, and this should be repeated until the weight is constant.

Finally, as a check upon the above, the crucible may be heated for twenty minutes over the blowpipe to wholly convert the calcic carbonate into calcic oxide ; after allowing to cool, weigh as $CaO$.

When there is such a small quantity of precipitate that it cannot be detached from the filter-paper, the residue, after ignition, may be advantageously treated with a little dilute hydrochloric acid and a few drops of pure sulphuric acid, then carefully evaporated to dryness, ignited, and weighed as $CaSO_4$.

2. If the lime is combined with phosphoric acid, treat the HCl-solution with ammonia until a precipitate begins to form, then just redissolve the precipitate in a few drops of HCl, and after adding an excess of a solution

of sodic acetate, precipitate the lime with ammonic oxalate.
The precipitate is further treated as in 1.

## B. Volumetrically.

(See **Alkalimetry**, also determination of hardness in
Water Analysis.)

## 2. Magnesia (MgO).

Magnesia is determined as—

a. **Magnesic pyrophosphate** $Mg_2 P_2 O_7$.

$$
\begin{array}{lll}
P_2O_5 & . \ 142 & . \quad 63{\cdot}96 \\
2MgO & . \ \ 80 & . \quad 36{\cdot}04 \\
\hline
& & 100{\cdot}00
\end{array}
$$

$$O = P\big\langle{}^{O}_{O}\big\rangle Mg$$
$$O = P\big\langle{}^{O}_{O}\big\rangle Mg$$

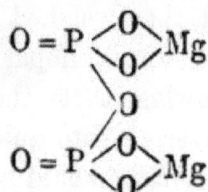

To the cold solution add $NH_4Cl$, and $NH_4(OH)$,
until it smells strongly of $NH_3$, then an excess of
**hydric disodic phosphate** $PO(OH)(NaO)_2$, well
stirring at the same time with a glass rod, but carefully avoiding contact with the sides of the beaker,
as otherwise the crystalline precipitate of ammonic
magnesic phosphate adheres to the glass. Allow
to stand in the cold for twelve hours. Filter and
wash on to the filter with a mixture of 1 part
strong ammonia and 3 of water, in which the precipitate is considerably less soluble than in water.
Continue washing with this mixture until a few
drops of the filtrate, after being acidulated with
nitric acid, give but a very slight opalescence with
argentic nitrate. Dry, ignite strongly in a platinum

crucible, and weigh as magnesic pyrophosphate of the composition given above.

*b.* **Pure magnesia** (MgO).—If the magnesia is combined only with organic acids, carbonic or nitric acids, the salt should be first gently ignited in a covered crucible, and finally strongly ignited and weighed as MgO.

### Separation of Lime from Magnesia.

*a.* To the hot solution containing sufficient $NH_4Cl$, add $NH_4(OH)$ and **ammonic oxalate** in excess, so that magnesic oxalate can also be formed. Allow the mixture to stand in a warm place for twelve hours, and then filter and treat the precipitate of **calcic oxalate** in the manner described on p. 29.

To the filtrate add $PO(OH)(ONa)_2$, precipitating the magnesia as **ammonic magnesic** phosphate, and weigh as pyrophosphate (see above).

If larger quantities of magnesia are present, the calcic oxalate precipitate often contains some magnesic oxalate, and in that case the clear supernatant liquid should be filtered first, whilst the precipitate is dissolved in HCl and reprecipitated with $NH_4(OH)$. After subsidence the filtration is continued as above, the magnesia being precipitated in the united filtrates.

*b.* **In the presence of phosphoric acid.**—To the HCl-solution add $NH_4(OH)$ until a large permanent precipitate forms, and redissolve this in acetic acid (the phosphates of lime and magnesia are soluble in both organic and mineral acids). In this acetic-acid-solution the lime is precipitated with ammonic oxalate, and the magnesia in the filtrate by means of $NH_4(OH)$ and $PO(OH)(ONa)_2$.

The separation of lime from the alkalies and magnesia is effected in the presence of $NH_4Cl$ by means of ammonic oxalate.

The separation of magnesia from the alkalies can be performed :—

a. By adding $Ba(OH)_2$ to a neutral solution free from all ammonic salts. Heat to boiling and filter. Dissolve the precipitate containing $Mg(OH)_2$ in HCl, and after separating the excess of baryta by means of dilute $SO_2(OH)_2$, precipitate the magnesia with $NH_4(OH)$ and $PO(OH)(ONa)_2$. The alkalies are separated from the baryta in the first filtrate by precipitating the latter with dilute $SO_2(OH)_2$.

b. If chlorides only are present : evaporate to dryness with pure oxalic acid and ignite. The ignited residue consists of magnesia and alkaline carbonates, which can be separated by extraction with hot water. Ignite and weigh the insoluble magnesia. Treat the solution containing the alkaline carbonates with HCl, evaporate to dryness, and weigh as alkaline chlorides, etc.

If sulphuric acid be present, it must first be removed with $BaCl_2$, and the filtrate is then evaporated with excess of oxalic acid as above. The magnesia will in this case contain some baric carbonate, which can be separated as in a.

III. Oxides of the Earthy Metals (Aluminium, Iron, and Manganese).

### 1. Alumina ($Al_2O_3$).

Aluminium is weighed as pure **alumina** ($Al_2O_3$), which is obtained by the ignition of the hydrate ($Al_2(OH)_6$) precipitated with ammonic hydrate.

The compounds of aluminium with volatile or organic acids can by ignition be also converted into pure alumina. The presence of non-volatile organic matters, such as tartaric acid, sugar, etc., prevent the complete precipitation with $NH_4(OH)$; they must therefore be previously destroyed by fusion with sodic carbonate and potassic nitrate.

The precipitation of alumina is effected in a hot and moderately-dilute solution by the addition of $NH_4Cl$ and $NH_4(OH)$ in *slight* excess. To complete the separation of the alumina, the liquid must be kept near the boiling-point until the free $NH_3$ has escaped; after subsidence the precipitate is filtered off and washed with boiling water. Dry, ignite in a platinum crucible, and weigh.

Alumina precipitated from a solution containing **sulphuric acid** carries down some of the latter with it, to render it free from which the precipitate should be dissolved in HCl and reprecipitated with $NH_4(OH)$ in the manner above described.

### 2. Ferric and Ferrous Oxides ($Fe_2O_3$ and FeO).

#### A. Gravimetric Determination.

1. **Ferric oxide** is usually precipitated as **ferric hydrate** $Fe_2(OH)_6$, which is then dried and converted

D

by ignition in a platinum crucible into the oxide ($Fe_2O_3$) as which it is weighed.

2. If both **ferric and ferrous oxides**, or only the latter is present in the HCl-solution, the ferrous oxide must first be converted into ferric oxide by heating with a little strong $NO_2(OH)$ before the precipitation with $NH_4(OH)$ can take place. If ferrous oxide only is present, then this can be calculated from the ferric oxide obtained ($Fe_2O_3 = 160$, equivalent to $2FeO = 144$).

The precipitation takes place by addition of $NH_4(OH)$, and then heating the liquid nearly to boiling. The precipitate must be well washed with boiling water until free from $NH_4Cl$, otherwise on ignition volatile ferric chloride ($Fe_2Cl_6$) would be formed, which would occasion loss.

When iron is to be separated from other metals it is often precipitated, by boiling the acetic-acid-solution, as **basic ferric acetate**, which, after filtration and washing with hot water containing a little ammonic acetate, is dried, strongly ignited in a platinum crucible, and weighed as $Fe_2O_3$.

3. **Determination of both $Fe_2O_3$ and FeO.** — If both oxides be present in HCl-solution, the rapid oxidation of the ferrous oxide can be prevented by adding a considerable quantity of $NH_4Cl$. This solution is then nearly neutralised with $NH_4(OH)$, so that the precipitate which at first forms is just redissolved, a strong solution of **sodic acetate** is now added, and the liquid boiled for a few minutes; the precipitated basic **ferric acetate** is washed by decantation with boiling water, and then converted into $Fe_2O_3$ as above described.

The colourless filtrate containing the ferrous acetate in

acetic-acid-solution is oxidised with strong $NO_2(OH)$ or $KClO_3$ and HCl, and the resulting ferric oxide precipitated with sodic acetate in the same way; and from the $Fe_2O_3$ obtained the FeO can be deduced by calculation.

## B. Volumetric Determination.

Iron, both as $Fe_2O_3$ and FeO, can be estimated by volumetric means.

### a. With potassic permanganate—

Potassic permanganate $(Mn_2O_6(KO)_2)$ dissolves in water, yielding a solution of such an intensely purple colour that 1 part in 1,000,000 of water is still easily recognisable by its red colour.

Dilute $SO_2(OH)_2$ decomposes the salt, liberating free permanganic acid, and so increasing the oxidising power. Concentrated HCl decomposes the salt with evolution of free chlorine; cold dilute HCl acts similarly, but more slowly. In order to prevent this decomposition, sulphuric-acid solutions should alone be used in titrations with potassic permanganate.

When permanganic acid $(Mn_2O_7)$ is brought in contact with oxidisable substances under favourable circumstances, it parts with $\frac{5}{7}$ of its oxygen, manganous oxide (MnO) being formed :—

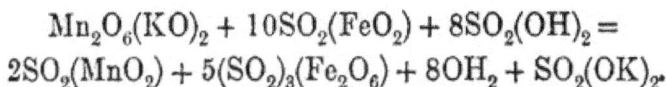

$$Mn_2O_6(KO)_2 + 10SO_2(FeO_2) + 8SO_2(OH)_2 =$$
$$2SO_2(MnO_2) + 5(SO_2)_3(Fe_2O_6) + 8OH_2 + SO_2(OK)_2.$$

### b. Standardisation of the potassic permanganate-solution :—

### a. By means of a solution of pure iron wire—

Dissolve 0·2 to 0·5 grm. of pure piano-wire in a small flask with pure dilute $SO_2(OH)_2$. The access of air to the flask should be prevented by placing a closed glass bulb in the mouth of the flask, which will thus act like a ball valve. Dilute the solution with water, and then run in the solution of potassic permanganate from a stoppered burette until a single drop imparts a permanent pink colour to the solution.

It is then easy to calculate what weight of iron 1 c.c. of permanganate-solution corresponds to. Thus, if 0·28 grm. of iron was dissolved, and for this 50 c.c. of permanganate-solution were required, then 1 c.c. permanganate = 0·0056 grm. iron.

### β. By means of ammonic-ferrous sulphate—

The crystallised salt $(SO_2(ONH_4)_2, SO_2(FeO_2), 6 \text{ aq.})$, can be readily obtained in the pure state; it contains $\frac{1}{7}$ of its weight of iron, and it is well suited for the standardisation of the permanganate-solution. Dissolve 1 to 2 grms. of the salt in cold water, add a little dilute $SO_2(OH)_2$, and titrate as above.

$$SO_2(ONH_4)_2 = 132$$
$$SO_2(FeO_2) \ = 152$$
$$6OH_2 \qquad = 108$$
$$\overline{\hphantom{000}392\hphantom{000}}$$

392 parts by weight thus contain 56 parts of iron, or $\frac{1}{7}$ of its weight.

Dissolve 1·96 grm. of the salt in a large quantity of

water, acidulate strongly with dilute HCl, and titrate with
the permanganate.    Then dilute the permanganate-solution
until 50 c.c. of it, equivalent to 0·28 grm. iron, are necessary
to effect the change of colour.    If 3·162 grms. of pure
potassic permanganate are dissolved in 1 litre of water,
then 100 c.c. of this solution = 0·56 grm. Fe = 0·72 grm.
FeO = 0·80 grm. $Fe_2O_3$.

### c. The analytical process—

Ferric oxide must be first converted into ferrous oxide.
Dissolve the substance in dilute $SO_2(OH)_2$, and then reduce
the solution by boiling with a little sodic sulphite, or by
introducing a small piece of pure zinc, and allowing the
latter to completely dissolve.    Potassic sulphocyanate must
give no red colouration with the solution, showing that
ferric salts are absent.

Now dilute with water and titrate, either the whole
solution or a definite volume of the same, with the permanganate.

If both ferric and ferrous oxides are present in the
same solution, the quantity of the latter can be determined
by means of permanganate; and then, by reducing the ferric
to ferrous oxide, the total iron may be ascertained, from
which the ferric oxide can be calculated by difference.

All substances capable of acting by reduction upon
potassic permanganate must be absent in the titration.

### d. With potassic dichromate—

The volumetric estimation of ferrous oxide by means of
a standard solution of potassic dichromate, offers several
advantages over the permanganate method, inasmuch as
the solution is more permanent, and thus requires less
frequent restandardisation; moreover, the dichromate can

be used in an ordinary burette with india-rubber-tube, and the presence of hydrochloric acid does not interfere with the process.   On the other hand, the end of the reaction between the ferrous oxide and the bichromate can only be ascertained by means of an external indicator, for which a drop of a freshly-prepared solution of potassic ferricyanide placed upon a white porcelain slab is used.  When a drop of the iron-solution is brought in contact with the ferri-cyanide, a rich blue colour is produced as long as the ferrous oxide is in considerable excess; but as the ferrous is more and more completely converted into ferric oxide through addition of bichromate, the blue acquires a more greenish tint; when this disappears the process is finished, and if an excess of bichromate be added, the indicator shows a brown colour.

The reaction between the ferrous oxide and the chromic acid may be thus expressed :—

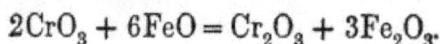

$$2CrO_3 + 6FeO = Cr_2O_3 + 3Fe_2O_3.$$

The reduction of ferric to ferrous compounds before titra-tion with the bichromate may be effected by means of sodic sulphite or zinc, in the same manner as described under the permanganate process.

PREPARATION OF THE DECINORMAL SOLUTION OF BI-CHROMATE.—4·917 grms. of potassic dichromate per litre. Since 1 molecule of potassic dichromate has three atoms of oxygen available for oxidising, a normal solution will contain $\frac{1}{6}$, and a decinormal solution $\frac{1}{60}$ molecular weight in grms. per litre.   But the molecular weight of potassic dichromate is 295, and thus the decinormal solution con-tains $295 \div 60 = 4\cdot917$ grms. per litre.

The standardisation is effected by means of steel wire or

ammonic-ferrous sulphate, in the same manner as in the case of the permanganate-solution.

### 3. Manganese.

The higher oxides of manganese present in soils, ashes, etc., are, on being heated with HCl, all converted into manganous chloride ($MnCl_2$) with evolution of chlorine gas.

The metal is determined as **manganous manganic oxide** ($Mn_3O_4$).

*a.* By precipitation as manganous $\left.\right\}$ $(CO(MnO_2))$.
     carbonate  .   .   .   .

Heat the solution of the manganous salt to boiling, and carefully add a freshly-prepared clear solution of sodic carbonate in slight excess. Allow the precipitate to subside, pass the supernatant liquid through a filter, and then boil up the precipitate in the beaker with water ; allow to subside, and then decant through the filter; repeat this washing by decantation three or four times, and then transfer the precipitate to the filter, and there wash thoroughly with boiling water. Dry, ignite strongly in a platinum crucible with the lid off in the oxidising flame. The filter should be separately ignited on platinum-wire and then dropped into the crucible. Weigh as $Mn_3O_4$.

*b.* By precipitation as hydrated **man-** $\left\{\begin{array}{l} MnO\,(OH) \\ MnO\,(OH) \end{array}\right.$
     **ganic peroxide**  .   .   .

The manganese should be in acetic-acid-solution (acetic acid keeps the alkaline earths and their phosphates in solution), and the precipitation is effected by means of **sodic hypochlorite, chlorine,** or, better still, a few drops

of bromine, the solution being maintained at a temperature of 60° C. for twenty-four hours. Filter hot, wash
well with hot water and dry; separate the precipitate
from the filter, ignite the latter, and dissolve both precipitate and ash in HCl. Evaporate off the greater part of the
HCl, dilute with water, and precipitate with sodic carbonate as in *a*.

*c.* By precipitation as **manganous** sulphide (MnS)—
Add ammonic chloride and ammonia to the manganous
solution contained in a flask until it shows a weak alkaline
reaction, then treat with yellow **ammonic** sulphide, fill
up the flask with water and allow to stand for twelve hours
in a moderately warm place. Wash the precipitate with
water containing ammonic sulphide, dissolve in HCl, and,
after boiling off the sulphuretted hydrogen, precipitate with
sodic carbonate as in *a*.

### 1. SEPARATION OF FERRIC OXIDE FROM ALUMINA.

*a.* **By sodic hydrate.**—The HCl-solution (if much free
acid be present, the greater part should be first neutralised
with sodic carbonate) is added drop by drop to a rather
concentrated boiling solution of pure sodic hydrate. The
iron is precipitated as ferric hydrate. Acidulate the filtrate with HCl, and precipitate the alumina with ammonia.
Redissolve the ferric hydrate (which always carries down
some soda with it) in HCl, and precipitate with ammonia.

*b.* **Volumetrically** and determination by difference.—
In the solution of a known weight of the two oxides in HCl
or dilute $SO_2(OH)_2$ determine the iron volumetrically by
means of permanganate or bichromate, and then calculate
the alumina by difference.

## 2. Separation of Ferric Oxide, Alumina, and Manganous Oxide.

### a. By ammonia in the presence of much ammonic chloride—

The dilute acid-solution containing much $NH_4Cl$ is heated to boiling, $NH_4(OH)$ is added in slight excess, and then volatilised again by continued heating; the iron and alumina are completely precipitated whilst the manganese remains in solution.

### b. By precipitation of the iron and alumina as basic acetates—

Dilute the HCl-solution and neutralise with sodic carbonate until the precipitate which forms is only just redissolved on shaking; heat to boiling, and add an excess of a strong solution of sodic acetate; continue the boiling for fifteen minutes—the precipitate contains the iron and alumina as basic acetates; allow to subside and filter, washing well by decantation. The filtrate contains the manganese, which can be precipitated with sodic hypochlorite, chlorine, or bromine, as in 3, *b*, on p. 39.

## 3. Separation of Alumina and Ferric Oxide from the Alkaline Earths.

By ammonia.—From the acid-solution the alumina and iron are precipitated by $NH_4(OH)$ in the presence of $NH_4Cl$, in the manner described under "determination of alumina." The filtrate contains lime and magnesia, which can be separated and determined in the manner already described.

4. SEPARATION OF FERRIC OXIDE, ALUMINA, MANGANOUS
   OXIDE, AND THE PHOSPHATES OF THE ALKALINE
   EARTHS.

*a.* By heating the acetic-acid-solution—

The precipitation is effected in an acetic-acid-solution
as described in 2, *b.* The precipitate contains iron, alumina, and phosphoric acid, but no other bases.

Dry, ignite and weigh the precipitate; dissolve it in
hot HCl, dilute and reduce the ferric oxide with a piece of
zinc, and determine the iron by titration with permanganate or bichromate.

The alumina and phosphoric acid are calculated by
difference; the phosphoric acid is determined in a special
portion of the solution according to II. 1, *a, a,* after the
addition of tartaric acid, or according to II. 1, *a, β.*

In the filtrate containing the alkaline earths and manganese, the latter is precipitated either—

   *a.* By heating with sodic hypochlorite, chlorine, or
   bromine.

   *β.* By ammonic sulphide; in this case the excess of
   ammonic sulphide must be decomposed with HCl,
   and the sulphur filtered off before the lime can
   be precipitated with ammonia and ammonic oxalate.

   In the filtrate from the lime remove the ammonia-salts
by evaporation to dryness and ignition, take up the residue
in HCl, and precipitate the magnesia with ammonia and
hydric disodic phosphate.

*b.* By means of **ammonic** sulphide—

Add $NH_4Cl$, and $NH_4(OH)$ to the solution until a precipitate just forms, then yellow ammonic sulphide, fill the

flask with water and cork it.    After standing in a warm place
for some time the precipitate (aluminic hydrate, ferrous and
manganous sulphides) separates; filter and wash with water
containing ammonic sulphide.    Dissolve the precipitate in
HCl, convert the ferrous into ferric oxide, and separate
this from the alumina and manganese as already described.
In the filtrate from the ammonic sulphide precipitate the
lime and magnesia are determined after destroying the
excess of ammonic sulphide with HCl and filtering off the
sulphur.

## II. ACIDS.

The following relates to the determination of acids combined with bases.    For the estimation of acids in the free
state see **Acidimetry, Alkalimetry.**

### 1. Phosphoric Acid.

Tribasic phosphoric acid, which is alone to be here
treated of, is determined as—

$\alpha$. Magnesic pyrophosphate.    $Mg_2P_2O_7$—

| | | | |
|---|---|---|---|
| $P_2O_5$ | . . . | 142 | . . . 63·96 |
| $2MgO$ | . . . | 80 | . . . 36·04 |
| | | | 100.00 |

$\beta$. Basic ferric phosphate.    $Fe_2(PO_4)_2$—

| | | | |
|---|---|---|---|
| $P_2O_5$ | . . . . | 142 | . . . 47·02 |
| $Fe_2O_3$ | . . . . | 160 | . . . 52·98 |
| | | | 100·00 |

$\gamma$. Volumetrically as uranic phosphate.

*u*. Determination as magnesic **pyrophosphate**—

    *a*. By **direct** precipitation with magnesia-mixture. —This method is applicable if the phosphoric acid be free or combined only with the alkalies. To the ammoniacal solution add magnesia-mixture,[1] stir well with a glass rod, but carefully avoid allowing it to touch the sides of the glass. Allow to stand ten hours in a covered beaker without heating. The precipitate of **ammonic** magnesic phosphate is further treated as described under "Determination of Magnesia."

    *β*. By precipitation as **ammonic molybdo-phosphate**, and subsequent determination as magnesic **pyrophosphate.**—This method is especially applicable in the case of solutions containing iron, alumina, or alkaline earths, combined with but little phosphoric acid (*e.g.* in soils). The phosphoric acid is precipitated by means of **ammonic molybdate**,[2] from a strong nitric-acid-solution. In the examination of superphosphates, etc., the solution for precipitation should not contain more than ·1 to ·2 grm. of phosphoric acid in 50 c.c.

    Treat the nitric-acid-solution in a beaker with such a quantity of ammonic molybdate that 50 parts of molybdic acid are added for every 1 part of phosphoric acid present. Keep for two to three

---

[1] Magnesia-mixture is prepared by dissolving 110 grms. of crystallised magnesic chloride, 140 grms. of ammonic chloride, and 700 c.c. of strong ammonia, in 1300 c.c. of water. 10 c.c. of this solution completely precipitate 0·24 grm. of phosphoric acid.

[2] The solution of *ammonic molybdate* is prepared by dissolving 1 part of the salt in 3 parts of ammonia (Sp. G. 0·96), and then adding 15 parts by weight of strong nitric acid (Sp. G. 1·2).

hours at 50° C., and then allow to stand six hours more covered up in a cold place. A portion of the clear liquid, on being treated with fresh ammonic molybdate, and heated to 60° C., should not give any further precipitate, and in case it does, more ammonic molybdate must be added to the whole, and allowed to stand again for ten hours.

Decant the clear solution through a filter, and wash the precipitate with a mixture containing 15 grms. of ammonic nitrate dissolved in 100 c.c. of water, together with 5 c.c. of strong nitric acid, and 10 c.c. of ammonic-molybdate-solution.

Dissolve the precipitate both on the filter and that remaining in the beaker in warm dilute ammonia, passing the whole through the filter; nearly neutralise the filtrate with HCl, and precipitate the phosphoric acid with magnesia-mixture. Filter, wash with ammonia (1 part strong ammonia to 3 parts of water), dry, ignite, and weigh as magnesic pyrophosphate.

*b.* Determination as **basic ferric phosphate**—

Applicable to solutions containing **much** phosphoric acid, and little iron and alkaline earths. The iron is precipitated with a portion of the phosphoric acid as $Fe_2(PO_4)_2$ by nearly neutralising the HCl-solution with $NH_4(OH)$, and then adding an excess of sodic acetate and acetic acid in the cold. Filter off the yellowish-white precipitate, wash repeatedly with hot water, dry, and ignite.

On treating the filtrate with ammonic oxalate, the **lime**

is precipitated.   Divide the filtrate from the lime into two parts; in the one determine the magnesia by ammonia and hydric disodic phosphate, and in the other the phosphoric acid by means of ammonia and magnesia-mixture.

c. **Volumetric** determination of combined phosphoric acid by means of **uranic acetate—**

Phosphoric acid forms with uranic oxide a compound of constant composition $Ur_4P_2O_{11}$, insoluble in acetic acid.

$$
\begin{array}{llll}
2UrO_3 & . \quad . \quad . \quad . \quad 576 & . \quad . \quad . & 80\cdot22 \\
P_2O_5 & . \quad . \quad . \quad . \quad 142 & . \quad . \quad . & 19\cdot78 \\
\end{array}
$$

$$\overline{\phantom{xxxxxxxx}100\cdot00}$$

Uranic acetate precipitates the whole phosphoric acid from a solution of ordinary phosphoric acid, or from an acetic-acid-solution of phosphates, provided no other acid but acetic is present.   An excess of the reagent can be identified by the reddish-brown precipitate of uranic ferro-cyanide, which a drop of the solution placed on a white slab gives when a drop of a solution of potassic ferrocyanide is brought in contact with it.   (It is necessary to employ a freshly-prepared solution of the ferrocyanide or the powdered salt itself.)

PREPARATION OF STANDARD URANIUM-SOLUTION.— About 35 grms. of crystallised uranic nitrate are dissolved in rather less than 1 litre of water.   The exact strength of this solution has now to be found by means of a standard solution of sodic phosphate.   For this purpose 23·1 grms. of pure crystallised hydric disodic phosphate, that has been previously finely powdered and dried by pressure between filter-paper, are dissolved in 1 litre of water.   In order to titrate the uranic-nitrate-solution with this standard sodic-phosphate-solution, an acetic-acid-solution of sodic acetate is required. This is prepared by dissolving 100 grms. of sodic acetate in water, adding 100 c.c. of strong acetic acid, and then making up to 1 litre with water.

ANALYTICAL PROCESS.—Measure 50 c.c. of the standard solution of sodic phosphate into a small beaker or flask, add ammonia until the solution is distinctly alkaline, then render acid with acetic acid, and add 5 c.c. of the sodic-acetate-solution. Heat to about 70° C., and run in the uranic nitrate very carefully from a burette, until a drop of the liquid taken out with a glass rod gives a **reddish-brown colour** with a freshly-prepared solution of potassic ferrocyanide placed upon a white tile.

The strength of the uranium-solution being thus ascertained, the latter is diluted with water until it corresponds exactly, cubic centimetre for cubic centimetre, with the standard sodic-phosphate-solution.

In the above titration of the uranic-solution with sodic phosphate it is really unnecessary to first add ammonia and then acetic acid ; but in the ordinary use of this process for the analysis of insoluble phosphates dissolved in mineral acids, it is essential to replace the mineral acid (usually nitric) by acetic acid. And in the standardisation of a solution all the conditions should be as similar to those in the actual analysis as possible. For the same reason the quantity of sodic acetate, as well as the bulk of the solution, should be kept as constant as possible.

This method is applicable even when the phosphate-solution contains a small quantity of iron (not more than about 1 per cent), but the presence of alumina interferes with the accuracy of the result. If iron be present, the addition of sodic acetate precipitates ferric phosphate, which must be filtered off, washed three or four times with boiling water, dried, ignited, and weighed as $(PO_4)_2Fe_2$. The filtrate is titrated with the uranium-solution in the usual way.

## 2. SULPHURIC ACID.

Sulphuric acid is always determined as **baric sulphate**.

| | | | | | |
|---|---|---|---|---|---|
| BaO | . . . | 153·0 | . . . | 65·67 |
| $SO_3$ | . . . | 80·0 | . . . | 34·33 |
| | | 233·0 | | 100·00 |

*a.* In sulphates **soluble** in water and hydrochloric acid—

The solution, which must be very dilute, is weakly acidulated with HCl, then heated nearly to boiling, and baric chloride added in slight excess. Allow to stand for twelve hours, then decant the clear liquid through a filter, wash the precipitate remaining in the beaker three or four times by decantation, boiling up with water each time; finally wash with boiling water on the filter until a few drops of the filtrate give no turbidity with sulphuric acid. Dry, transfer the precipitate to a platinum crucible, ignite the filter on platinum-wire, and treat the ash with a drop of hydrochloric and sulphuric acids in the lid of the crucible, evaporate off the acid fumes very carefully, and then ignite the crucible with the lid on, weigh.

If the original solution contains nitrates, these must be destroyed by evaporation with HCl; and if much free HCl be present, this must be evaporated or nearly neutralised.

*b.* In sulphates insoluble or difficultly **soluble** in

water or hydrochloric acid.   (Sulphates of barium, strontium, and calcium.)

a. In the **dry way**—

Fuse the finely-powdered substance with four or five times its weight of fusion-mixture $CO(ONa)_2$, $CO(OK)_2$ **(free from** sulphates**)** in a platinum crucible over the blowpipe for ten minutes.   Extract the fused mass, after cooling, with boiling water in a beaker.   **The alkaline sulphates and carbonates pass into solution whilst the carbonates of the alkaline earths remain undissolved.**

Filter, wash the insoluble residue with water containing a little ammonia and ammonic carbonate.   Acidulate the filtrate, which contains the whole of the sulphuric acid, with HCl, and precipitate with baric chloride as described in a.

β. In the **wet way**—

The above insoluble sulphates can also be decomposed in the wet way.   Whilst calcic and strontic sulphates are decomposed by six hours' digestion in the cold with a strong solution of ammonic carbonate and ammonia, baric sulphate, on the other hand, remains unacted upon.   Baric sulphate, however, can be completely decomposed by **repeated boiling** with a solution of potassic carbonate, and pouring off the clear liquid.   Strontic and calcic sulphates are readily decomposed by boiling with a solution of sodic or potassic carbonates. In every case soluble alkaline sulphates are formed and pass into solution, whilst the insoluble carbonates of the alkaline earths remain behind.

### 3. Silicic Acid.

Silicic acid is always weighed in the anhydrous state after ignition.

The decomposition of silicates has already been treated of in detail (p. 11).

The acid-solutions (together with any silica that may have separated) are in all cases evaporated to dryness in a porcelain dish over a water-bath, and the residue heated to complete dryness, until no more acid fumes are given off. The final heating is most advantageously effected by placing the porcelain dish on wire-gauze over a bunsen ; the heat must not be excessive, or else insoluble silicates may be formed by the combination of the bases with the silica.

After cooling, the mass is moistened with HCl, and allowed to stand for half-an-hour, then it is heated on the water-bath and diluted with water. After subsidence filter, wash well with hot water, dry, and ignite in a platinum crucible very carefully, lest some of the light silica is carried away. Weigh as $SiO_2$.

The purity of the silicic anhydride can be tested by digesting a portion of it for one hour with a moderately-concentrated solution of sodic carbonate (6 c.c. of a saturated solution of $CO(ONa)_2$, with 12 c.c. of $OH_2$ for every ·1 grm. $SiO_2$) in a platinum dish on the water-bath. The whole of the silica, if pure, should dissolve.

## 4. HYDROCHLORIC ACID.

Always determined in the form of **argentic chloride**, AgCl.

| | | | | | |
|---|---|---|---|---|---|
| Ag | . | . | 107·97 | . | . 75·28 |
| Cl | . | . | 35·46 | . | . 24·72 |
| | | | 143·43 | | 100·00 |

**A. Gravimetric determination.**—Acidulate the cold diluted solution with a little nitric acid, and then add argentic nitrate-solution in slight excess. Heat to boiling, and stir vigorously with a glass rod until the precipitated argentic chloride becomes aggregated into flakes. Allow the liquid to become quite clear, and then pass the super-natant fluid through a filter; wash the precipitate three times by decantation, boiling up each time with water acidulated with nitric acid ; transfer to the filter, and wash repeatedly with boiling acidulated water, until a few drops of the filtrate give no turbidity with dilute hydrochloric acid ; finally wash once or twice with ordinary distilled water to remove the acid from the filter-paper, which would otherwise fall to pieces on drying. Dry thoroughly, transfer the precipitate to a porcelain crucible, ignite the filter-paper on the lid of the crucible, and treat the ash with a drop of nitric and hydrochloric acids. Evaporate off the acid very cautiously, and then ignite the crucible with the lid on, until the precipitate begins to fuse round the edges. Allow to cool, and weigh as AgCl.

**B. Volumetric determination.**—A standard solution of argentic nitrate can be employed to determine chlorides soluble in water

Prepare a perfectly **neutral** solution of argentic nitrate, containing $\frac{1}{10}$ molecular weight of $NO_2(AgO)$ in grms. = 17 grms. per litre.

Dissolve several separate portions of pure ignited sodic chloride, each weighing ·10 to ·15 grm. in 20 to 30 c.c. of distilled water each, and add to each 4 or 5 drops of a cold saturated solution of yellow potassic chromate.

Now run in the decinormal solution of argentic nitrate very carefully from a burette, until the last drop or two produce a permanent blood-red **tinge**. Until the whole of the chloride present has been precipitated, no red colour is produced; but the first drop of argentic nitrate in excess is at once rendered apparent by the formation of red argentic chromate.

From this titration the quantity of argentic-nitrate-solution required for $\frac{1}{10}$ molec. wt. of NaCl in grms., = 5·846 grms., can be calculated, as well as the quantity of water that must be added to 1000 c.c. of the argentic-nitrate-solution to make it of the required strength.

Acid-solutions must be **exactly neutralised** before titration.

1000 c.c. argentic - nitrate - solution = $\frac{1}{10}$ molec. wt. HCl = 3·646 = chlorine 3·546 = 5·846 NaCl = 7·46 KCl, etc.

## 5. Nitric Acid.

The following are the methods which are most frequently employed in the estimation of combined nitric acid :—

   *a.* If the bases are only combined with nitric acid, the latter can be determined **by loss** after all the bases have been estimated.

   *b.* By **fusing** the nitrate with a weighed quantity of

potassic dichromate, borax, or silica, and then calculating from the **loss in weight**. This method is applicable to the estimation of nitric acid in Chili saltpetre. (See the Analysis of Manures—Chili Saltpetre.)

c. By **distillation** of the nitrate (1 to 2 grms.) with a mixture of sulphuric acid and water (1 vol. of $SO_2(OH)_2$ to 10 of $OH_2$). The mixture is heated for three or four hours in a tubulated retort on a sand-bath to a temperature not exceeding 170° C. The distillate is collected in a flask containing a measured quantity of normal caustic soda. The soda left unneutralised is determined volumetrically (see Alkalimetry), and the nitric acid calculated from the difference.

d. **By decomposition of the nitric acid with ferrous chloride—**

　a. **Pelouze's method.**—Heat a weighed quantity of the nitrate with a strongly acid solution of ferrous chloride of known strength, and determine, after the reaction is completed, the quantity of iron still present in the ferrous condition; the difference represents the iron that has suffered oxidation, from which the nitric acid can be calculated :—

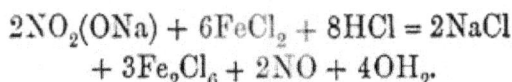

$$2NO_2(ONa) + 6FeCl_2 + 8HCl = 2NaCl + 3Fe_2Cl_6 + 2NO + 4OH_2.$$

From the above equation it appears that 6Fe = 336 parts by weight of Fe converted into ferric oxide, are equivalent to $N_2O_5 = 108$ parts by weight of nitric anhydride.

β. **Schlösing's** method. — This method is based upon the same reaction as Pelouze's, but instead of the ferrous chloride oxidised, the **nitric oxide** evolved is made the measure of the nitric acid present—

As this method is the one which, under ordinary circumstances, is the best adapted for the determination of nitric acid, it will be here described at some length; whilst the modified process devised by **Warington** for the estimation of nitrates in soils, etc., will be described in detail in the Analysis of Soils.

The solution of the nitrate is placed in a small tubulated retort, the neck of which is cut short and contains a doubly perforated india-rubber-cork. Through one of the perforations passes a small glass-tube, connected above with a little funnel by means of a small piece of india-rubber tubing fitted with a clamp. Through this funnel the solution of the nitrate, the concentrated ferrous chloride, the hydrochloric acid, and a little water, can be successively passed. The ferrous chloride and the acid are first poured into the glass vessel which contained the nitrate solution, and thus serve to rinse it out. Through the other perforation of the cork in the neck of the retort passes a delivery-tube leading to a mercury-trough. After the liquids have been duly introduced through the funnel, as above described, the clamp attached to the latter is closed, and a stream of carbonic anhydride from a constant generator is passed by means of a tube fitted with a cock, passing through a cork in the tubulure of the retort. By means of this stream of carbonic anhydride the whole of the air is expelled from the apparatus, which is indicated by a few bubbles of the escaping gas collected in the mercury-trough being completely absorbed by a strong solution of caustic potash. The cock of the carbonic acid apparatus must now be closed, a large test-tube, drawn out to a point at the closed extremity and filled with mercury and a little milk of lime, is placed over the delivery-tube, and the retort is heated to decompose the nitrate with the ferrous chloride. Seven or eight minutes are sufficient to complete the reaction. Whilst the reaction is going on the cock of the carbonic acid apparatus should be slightly opened several times, and when the reaction is over, a stream of carbonic anhydride is passed through until the whole of the nitric oxide evolved has been driven into the test-tube placed for its reception in the mercury-trough. (The tube for collecting the gas in is most suitably made by sealing, in the blowpipe, the narrow end of a retort " adaptor.")

The gas in the tube is shaken up with the milk of lime to remove all carbonic anhydride and acid vapours from the nitric oxide.

The nitric oxide has now to be reconverted into nitric acid, which can then be estimated volumetrically by means of standard alkali. For this purpose the closed end of the tube containing the gas is connected by means of india-rubber and glass tubing with the drawn-out neck of a flask containing a little water. This water is first boiled to expel the air from the flask and tube, and whilst boiling the tube must be attached to the small extremity of the vessel containing the gas. The point of this vessel is now broken off carefully after the connection has been made. At first aqueous vapour from the flask rushes into the gas-tube, but very soon the condensation of the vapour in the flask draws the gas into the latter, the rate of absorption being regulated by compressing the india-rubber-tube with

the finger.   As soon as the milk of lime rises to the level
of where the india-rubber-tube is attached, the latter must
be at once clamped.

The nitric oxide is now contained partly in the flask and
partly in the tubing connecting it with the gas-tube, together with a few bubbles at the top of the gas-tube ; to
drive these into the flask also 20 to 30 c.c. of hydrogen are
made to enter the gas-tube from below, and by opening
the clamp this also is drawn into the flask.   The clamp is
again closed and the gas-tube detached.   A gas-holder containing oxygen is now connected by means of the india-
rubber-tube with the flask, and on opening the clamp the
oxygen is allowed to enter.   The clamp is now again closed,
and the flask left to stand for a quarter of an hour until
all the nitrous fumes have been absorbed.

The acid formed has then only to be titrated with a
standard solution of lime.

**Schlösing's method   (by measurement of the
nitric oxide evolved)—**

Instead of first converting the nitric oxide evolved in
the reaction with ferrous chloride, the volume of the gas
may be at once determined, and from this the nitric acid
calculated.

The solutions of ferrous chloride and hydrochloric acid
are heated to boiling in a small flask fitted with a stoppered
funnel and a delivery-tube ending in a very small aperture
in a pneumatic trough filled with water.   The boiling is
continued until all the air is displaced from the flask, and
then the solution of the nitrate is slowly added by means
of the stoppered funnel, the liquid adhering to which is
rinsed down with hydrochloric acid.   The gas evolved is
collected in a measuring-tube placed over the end of the

delivery-tube in the pneumatic trough. The reaction is known to be complete when the vapours condensing in the delivery-tube form an unbroken column uninterrupted by bubbles of gas.

As soon as the operation is over a new measuring-tube should be placed over the delivery-tube, and a fresh experiment commenced whilst the liquid in the flask is still boiling.

In order to obviate the trouble of making a correction for the temperature and pressure at which the gas is measured, and, moreover, to avoid the errors arising from the slight solubility of the nitric oxide in water and the disturbing influence of any oxygen dissolved in the water, the following plan may be adopted with great advantage :— A determination is first made with a known weight of a pure nitrate, and the quantity of gas obtained from this compared with the quantity yielded by the same weight of the nitrate under examination. The two volumes of nitric oxide, being under the same conditions of temperature and pressure, are proportional to the weights of nitric acid from which the gases were respectively obtained.

In this manner as many as ten nitric acid determinations can be made without changing the ferrous chloride or the volume of nitric oxide for comparison.

The method is especially applicable to the analysis of manures rich in nitrates, such as saltpetre and Chili saltpetre. If standard solutions of equal strength, both of the pure nitrate and of the nitrate under examination, be used (using a solution of pure potassic nitrate to compare with ordinary saltpetre, and pure sodic nitrate for comparison with Chili saltpetre), then the ratio of the volume of gas obtained from the substance in question to the volume of gas obtained from the pure substance, expresses the percentage of the pure nitrate in the former. Thus, for example, if the ratio in the case of Chili saltpetre be 0·86, the latter contains 86 % of pure sodic nitrate.

   *c.* **By conversion of the nitric acid into ammonia—**

      *a.* The reduction is effected by means of nascent hydrogen in an alkaline solution. The ammonia formed is distilled into a measured quantity of standard acid, the excess of which is then determined by means of standard alkali.

      A mixture of zinc and iron filings, when boiled with a moderately strong solution of caustic potash, produces an evolution of hydrogen; if a nitrate is added to this mixture ammonia is formed:—

$$NO_2(OH) + 4Zn + OH_2 = NH_3 + ZnO.$$

      The process is conducted by using a mixture of 50 grms. of granulated zinc and 25 grms. of iron filings, which have been purified by sifting and ignition in a closed vessel. To this is added the nitrate (·5 grms.), 20 c.c. of water, and 20 c.c. of caustic potash of 1·3 Sp. G.

      From the above equation it is apparent that 17 parts by weight of ammonia are equivalent to 54 parts of nitric acid.

      The residuary zinc and iron can, after being washed with water and dilute acid, be used again for the same purpose.

      Instead of the above proportions, 1 grm. of

the nitrate may be mixed with 4 grms. of iron and 10 grms. of zinc filings, 16 grms. solid potassic hydrate, and 100 c.c. of alcohol (Sp. G. ·825). The alcohol diminishes the danger of the liquid boiling over.

[See also Aluminium Method of determining Nitrates, in "Water Analysis," p. 275.]

β. Instead of the above, a spiral of zinc and iron-foil soldered together may be introduced into the alkaline liquid containing the nitrate. The nascent hydrogen reduces the nitric acid to ammonia as before, which is then determined as **nitrogen gas** by means of **sodic hypobromite** in the **azotometer** (W. Wolf, *Chem. Centralblatt*, 1863, 651). (See p. 25.)

γ. The reduction may be effected in **either acid or alkaline solution** by means of nascent hydrogen, the ammonia formed being volumetrically determined by decomposition with **brominated sodic hypochlorite.** (Krocker and Dietrich, *Zeitschrift f. anal. Chemie*, iii. 162, v. 36 ; and Wagner, *id.* xiii. 383.)

The nitrate is reduced by means of zinc and sulphuric acid in a flask fitted with a cork bearing a tube containing glass beads moistened with dilute sulphuric acid to absorb any escaping ammonia. After the action has continued for several hours, the contents of the flask, together with the glass beads, are washed into a beaker, sodic carbonate is added until the reaction is strongly alkaline, and then the ammonia is determined by decomposition with

brominated sodic hypochlorite, the strength of
which has been determined with a solution
of arsenious acid. (See Determination of
Ammonia, p. 23.)

δ. The methods for determining nitric acid in
water will be described under the heading,
"Water Analysis."

## 6. Carbonic Acid.

The determination of free carbonic acid in water will be
treated of under "Water Analysis."

The determination of combined carbonic acid is generally
effected by one of the following
methods :—

a. By the **loss in weight**
suffered on the carbonate
being decomposed with a
**stronger acid**—

The most convenient form
of apparatus devised for this
mode of determination is
Schrötter's (Fig. 2), which
consists of a thin glass-bulb
possessing a stoppered aperture at the side, and bearing two enlarged tubes at
the top. One of these (A)
is furnished with a stopcock
below, and with a stopper
above; it serves to contain
the acid with which the carbonate is decomposed.

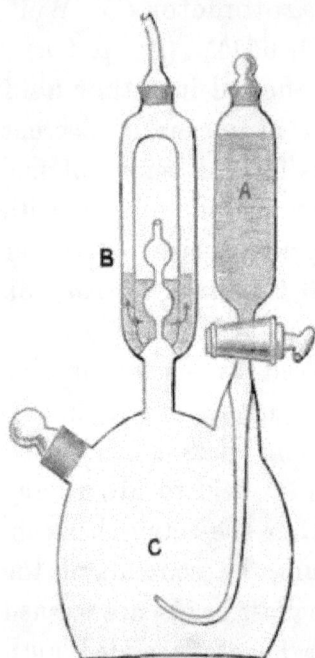

Fig. 2.

The other (B) is so arranged that a little strong sulphuric acid can be placed in it, through which the carbonic anhydride evolved in the bulb can bubble, and so be deprived of any moisture that it may carry with it. Above, this tube ends in a nozzle, which is ground in like a stopper.

Before the apparatus is used the one tube (A) must be filled with dilute nitric or sulphuric acid ; the other (B) must be half filled with strong sulphuric acid. The whole apparatus is now to be weighed, and then again, after the carbonate (about 1 grm.) has been introduced into the bulb through the stoppered aperture, the increase in weight denotes the quantity of substance taken. The stopper of the tube (A) is now removed, and the dilute acid allowed to enter gradually by opening the stopcock below. The carbonate is decomposed with effervescence, and the evolved carbonic anhydride passes up through the sulphuric acid in the tube (B), where it is deprived of any moisture. The decomposition of the carbonate must be complete before the whole of the acid has been added, to ensure that sufficient acid has been employed. When the whole of the acid has been added, the apparatus should be placed upon a sand-bath and heated to gentle ebullition ; at the same time a current of dried air should be aspirated through the apparatus by fitting an india-rubber-tube to the nozzle of the tube (B), and allowing the dried air to enter by the top of the tube (A). The aspiration can be most conveniently effected by carefully sucking with the mouth at the end of the tube attached to the nozzle of B. The

apparatus is then allowed to again acquire the temperature of the air, which it should do in about an
hour's time.    After the outside has been carefully
wiped with a handkerchief, it is weighed, and the
loss in weight represents the carbonic anhydride
evolved from the quantity of carbonate taken.

Fig. 3 shows a form of apparatus constructed
upon the same principle as
Schrötter's.    It is due to the late
Mr. Valentin.    The bulb A contains the acid with which the
carbonate placed in C is to be
decomposed.    This acid can be
admitted drop by drop into C,
when the india-rubber-stopper(D)
is removed.    The extremity of
the delivery-tube (E) is curled
and drawn out to a point in order
to prevent the acid from running
out before the stopper (D) is removed.    The evolved carbonic anhydride is dried
in passing through the bulbs (B), which contain
sulphuric acid.    The flask (C) should only have a
capacity of 1 to 2 oz.

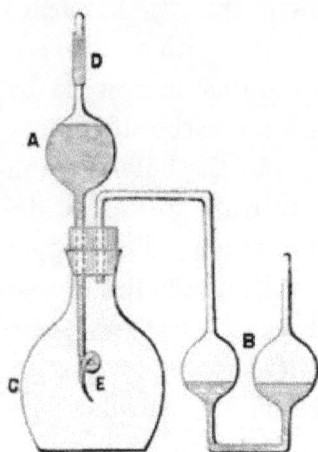

Fig. 3.

b. By determining the **gain in weight** of an **absorption-tube**—

A weighed quantity of carbonate (about 1 grm.)
is placed in a small flask fitted with an india-rubber-
cork bearing a stoppered funnel and a delivery-tube.
The latter is connected with a series of U-tubes, the
first of which is empty, and serves to condense any
moisture that may pass over from the decomposing

flask ; the second contains pumice soaked in strong
sulphuric acid to completely stop any moisture in
the evolved carbonic anhydride ; and the third contains anhydrous cupric sulphate to arrest any vapour
of hydrochloric acid that may be still present in the
gas.    The third U-tube is further connected with
one end of a set of Liebig's or Geissler's potash-
bulbs (see p. 79) containing strong caustic potash-
solution (1 of KOH to 1 of $OH_2$); to the other
extremity of these bulbs is attached a small bulb
containing a few drops of strong sulphuric acid to
stop any moisture that may be carried over from
the caustic potash.

The weight of the potash-bulbs with the little
sulphuric-acid-bulb is taken before the experiment.
The carbonate is decomposed in the flask by allowing dilute hydrochloric acid to enter through the
stoppered funnel.    When the decomposition is complete, the flask is heated to gentle ebullition on a
sand-bath, a current of air that has been freed from
carbonic anhydride being aspirated through the
apparatus by attaching an india-rubber-tube from
the aspirator (for which an inverted washbottle may
be conveniently used) to the open end of the little
sulphuric-acid-bulb.    The air enters through the
top of the stoppered funnel.    When the whole of
the gas in the apparatus has been drawn through,
the potash-bulbs, with the little sulphuric-acid-bulb,
are disconnected, and after standing for about twenty
minutes they are again weighed.    The increase in
weight denotes the quantity of carbonic anhydride
yielded by the weight of carbonate employed.

c. By the **loss in weight** suffered on fusing **anhydrous** compounds with four times their weight of borax.

d. By **determination of the alkaline base** in a neutral carbonate of an alkali or alkaline earth by saturation with a standard acid. (See Alkalimetry.)

e. By **gasometric-determination** of the carbonic anhydride evolved on decomposition of the carbonate with hydrochloric acid at the ordinary temperature.

# III. DETERMINATION OF FREE ACID OR ALKALI IN SOLUTION.

## (ACIDIMETRY AND ALKALIMETRY.)

### A. By means of the Specific Gravity.

IF the acid or alkali to be estimated is free from all other dissolved matters, its quantity can be determined by taking the specific gravity of the liquid, since the relation of the specific gravity to the percentage strength of aqueous solutions of the more important acids and alkalies has been accurately determined by experiment. (See tables in the Appendix.)

### B. By Saturation of the free Acid or Alkali.

By means of standard solutions of alkali in the first, and of acid in the second case.

The following test-solutions are required :—

1. An acid-solution of known strength.—**Normal acid,** sulphuric, oxalic, or nitric acid is used for the purpose, in such a state of dilution that 1000 c.c. contain the molecular weight (or half the molecular weight) in grms. of the acid. Thus :—

|  |  |  |  |  | $\frac{1}{2}$ Molec. Weight. |
|---|---|---|---|---|---|
| 1000 c.c. normal sulphuric acid | . | . | 40 | grms. | $SO_3$ |
| ,, | ,, | oxalic | ,, | . . | 63 ,, $C_2H_2O_4 + 2OH_2$ |
| ,, | ,, | nitric | ,, | . . | 54 ,, $N_2O_5$. |

F

2. The solution of an alkali of known strength.—**Normal alkali** :—Caustic soda is the alkali chiefly used ; it is of such strength that—

1000 c.c. of normal soda ($\frac{1}{2}$ molec. wt.) $= 31$ grms. $Na_2O$. Thus any given volume of this soda-solution will exactly neutralise the same volume of any of the above normal acids.

## Preparation of the Normal Acid.

a. **Normal sulphuric acid**—

Pour about 60 grms. of strong sulphuric acid into a flask containing about 1020 c.c. of distilled water, mix well, and allow the mixture to cool to about $18°$ C, and then determine the sulphuric acid in several measured portions, of 20 c.c. each, by means of baric chloride.

Taking the mean of these determinations, dilute with water until 1000 c.c. contain 40 grms. of $SO_3 = \frac{1}{2}$ molec. wt.

Thus, if ·88 grm. $SO_3$ was found in 20 c.c., then 1000 c.c. contain 44 grms. of $SO_3$ ; therefore $(40 : 1000 :: 44 : x = 1100)$ every 1000 c.c. of the trial-solution must be diluted to 1100 c.c. Fill a litre-flask exactly with the acid-solution, and put the requisite quantity of distilled water (100 c.c.) into a perfectly dry larger flask or bottle ; then pour the acid from the litre-flask into the latter, and wash out the former by pouring back the solution several times.

b. Normal **oxalic acid**—

Dissolve $\frac{1}{2}$ molec. wt. in grms. $= 63$ grms. of pure crystallised oxalic acid, $\begin{cases} CO(HO) \\ CO(HO) \end{cases}$, $2OH_2$, in distilled water, and dilute to 1000 c.c.

Commercial oxalic acid, which generally contains calcic and hydric potassic oxalates, should be purified by dissolving in six to eight times its weight of water, rejecting the first crop of crystals, and purifying the deposit from the mother-liquor by repeated recrystallisation, and then finally thoroughly drying the crystals by exposure to the air.

c. **Normal nitric acid**

is used in those cases in which sulphuric and oxalic acids would form insoluble compounds, as in the saturation of the caustic and carbonated alkaline earths.

Dilute nitric acid is rendered normal—

α. By **normal caustic soda**, the nitric acid being diluted until 1 c.c. of the acid is exactly saturated by 1 c.c. of the soda; then 1000 c.c. of the acid so diluted will contain 54 grms. of $N_2O_5 = \frac{1}{2}$ molecular weight in grms.

β. By **calcic carbonate.**—Pure pieces of Iceland Spar are best adapted for the purpose. Some pure dilute nitric acid is first further diluted until it corresponds, volume for volume, with a pure solution of alkali.

1 grm. of pure calcic carbonate is placed in a flask; to this an excess of the acid, prepared as above, is added, and the carbonic anhydride is expelled by heating. The excess of acid is now titrated with the alkali, and the number of c.c. of acid neutralised by the 1 grm. of calcic carbonic thus ascertained.

Thus, if 10 c.c. of acid were employed, and 3 c.c. of the alkali were requisite to saturate the excess of acid, then the 1 grm. of $CO(CaO_2)$ was neutralised by 7 c.c. of acid, or ·5 grm. by 3·5 c.c. of acid.

But $\frac{1}{2}$ molec. wt. of $CO(CaO_2)$ in grms. = 50 grms., therefore 350 c.c. of the acid must be diluted to 1000 c.c., in order to obtain a normal acid, of which 1 c.c. saturates ·028 grm. CaO, or ·031 grm. $Na_2O$, etc.

### Preparation of the Normal Alkali.

By means of the test-acid an alkaline-solution—generally a solution of **caustic soda**—is prepared, which in 1000 c.c. contains $\frac{1}{2}$ molec. wt. in grms. = 31 grms. $Na_2O$.

For the accurate determination of small quantities of acid it is preferable to employ a solution containing ·0155 grms. $Na_2O$ per 1000 c.c.

A freshly-prepared solution of caustic soda (free from carbonate) is diluted until it has a specific gravity of about 1·05. Now measure 25 c.c. of normal acid into a beaker, and tint it red with litmus, then carefully allow the diluted soda-solution to run into the beaker from a burette with a pinch-cock, until the red colour of the liquid is changed to blue. The soda-solution is then further diluted until 25 c.c. of it exactly saturate 25 c.c. of the normal acid.

The vessel in which the normal alkali is preserved should be fitted with a bored cork, bearing a small tube filled with sodic sulphate and caustic lime, to exclude the atmospheric carbonic anhydride.

In order to remove the carbonate contained in commercial caustic soda, the solution of the latter is heated to boiling in a flask, some milk of lime is added, and the flask closed with a cork provided with a tube as described above.  After subsidence, the clear solution is titrated as above.

## Preparation of the Indicator.

1. **Litmus.**—Mix some litmus with six times its weight of distilled water in a flask.  After digestion and subsidence, pour off the clear liquid and divide it into two equal parts.  To the one part add nitric acid until a red colour is obtained, then mix the other part with it, and add one part of strong alcohol to the whole.

Instead of adding nitric acid, the litmus may with advantage be treated with some freshly-precipitated gypsum.

Alkalies impart to the litmus-solution a **blue**, carbonic acid a **violet**, and stronger acids a bright **red**, colour.

As the violet colour, due to carbonic acid, makes its appearance in a cold solution before the whole of the alkaline carbonate has been completely saturated with the stronger acid, it is necessary to heat solutions containing carbonates to boiling whilst the neutralisation is going on ; the acid added may also be coloured red.

` N.B.—Wartha (*Berlin Berichte*, ix. 217) renders the litmus more sensitive by treating the commercial litmus with methylated spirit in the cold ; the residue is then digested for twenty-four hours with distilled water, and the solution evaporated to dryness on the water-bath.  The extract is then repeatedly treated with absolute alcohol containing a little acetic acid, and finally evaporated to dryness.  A crisp residue remains behind, which forms a brown powder ; absolute alcohol containing acetic acid removes from this a large quantity of a scarlet colouring-matter, which does not turn blue with ammonia.  The residue is dissolved in water, evaporated to dryness on the water-bath, and by repeatedly moistening with absolute alcohol and evaporating, the whole of the acetic acid is removed. The powdered colouring-matter so obtained can be preserved for use.

2. **Cochenille** possesses the advantage of being unacted upon by free carbonic acid. The tincture is brownish-red, and it becomes orange-yellow with acids, rose-red with alkalies. It cannot be used in the presence of salts of iron.

In preparing for use, 1·5 grms. of powdered cochenille are digested with a mixture of 5 c.c. of alcohol and 200 c.c. of water; the liquid is frequently shaken up, and finally filtered.

3. **Coralline** dissolved in alcohol forms a very sensitive indicator, especially for colourless liquids. Free acid produces a yellow colour, whilst the least excess of alkali gives rise to a distinctly red tint.

## The Analytical Process.

### 1. Determination of the Acidity of a Solution.

By using the test-alkali the quantity of any acid can be readily determined.

1000 c.c. of the normal alkali ($= 31Na_2O$) correspond to $\frac{1}{2}$ molec. wt., expressed in grms. of every oxy-acid; they therefore neutralise—

40 grms. of sulphuric anhydride.

| | | | |
|---|---|---|---|
| 66 | „ | anhydrous tartaric acid. | |
| 75 | „ | crystallised | „  „ |
| 63 | „ | „ | oxalic acid. |
| 58 | „ | anhydrous citric | „ |
| 67 | „ | crystallised „ | „ |

51 grms. of acetic acid.

60    „    hydrous acetic acid.

188    „    acid potassic tartrate.

54    „    nitric acid.

If, for example, 5 grms. of nitric acid, the strength of which is to be ascertained, be taken, and 20 c.c. of normal caustic soda be used to saturate it, then $(1000 : 54 : : 20 : x)$ 1·08 grms. of $N_2O_5$ would be contained in 5 grms., or the acid would be of 21·6 % strength.

Since 100 c.c. of normal caustic soda always saturate $\frac{1}{20}$ of the molec. wt. in grms. of the pure acid in question, therefore, if $\frac{1}{20}$ of the molec. wt. in grms. of the acid-solution under examination be taken, the number of c.c. of normal alkali used will at once express the percentage strength of the liquid. Since larger quantities are more easily weighed, it is preferable to take $\frac{1}{4}$ of the molec. wt. in grms. of the acid, to dilute this with water to 500 c.c., and then to use 100 c.c. $= \frac{1}{20}$ of the molec. wt. for the experiment.

## 2. Determination of a free or carbonated Alkali.

By employing a test-acid the quantity of any alkali can be readily detected.

1000 c.c. of the normal acid (containing $\frac{1}{2}$ molec. wt. in grms.) saturate—

| | | | | |
|---|---|---|---|---|
| Potash | . . . 47·11 grms. | Ammon. carbonate | . 48 grms. |
| Potassic hydrate | . 56·11 „ | Soda | . . . . 31 „ |
| Potassic carbonate | 69·11 „ | Sodic hydrate | . . 40 „ |
| Crystallised sodic | | Sodic carbonate | . 53 „ |
|    carbonate | . . 143 „ | Lime | . . . . 28 „ |
| Ammonia | . . 17 „ | Calcic hydrate | . . 37 „ |
| Ammonic oxide | . 26 „ | Calcic carbonate | . 50 „ |

The following points must be attended to, according to the nature of the substance under examination :—

a. To ascertain the total alkalinity, irrespective of its being caustic or carbonated, the solution should be supersaturated with a measured quantity of standard acid in a little flask, so that after boiling (if any carbonate be present) the litmus with which the solution was coloured at first, becomes of a permanently red colour. The excess of acid added is then titrated with normal soda, until the blue colour of the litmus just returns. This method of titrating the excess of reagent is both more expeditious and accurate. The quantity of acid requisite for saturation, and the quantity of alkali corresponding to this, can then be easily calculated.

b. If in the case of alkaline carbonates $\frac{1}{20}$ of the molec. wt. in grms.,—thus, 5·3 grms. dry sodic carbonate, and 6·9 grms. potassic carbonate,—be used, and the process carried out as in (a), then the number of c.c. of the normal acid used express at once the percentage of the pure salt in the substance examined.

If tincture of cochenille be used as indicator, it is unnecessary to boil the liquid.

c. In using the process for the estimation of alkali and alkaline carbonate in crude potash or the ashes of wood, etc., it is necessary to first separate the salts insoluble in water by filtration, as they would influence the result. The chloride and neutral sulphate of potassium are without interfering action, whilst the alkaline silicates and phosphates, possessing an alkaline reaction, are saturated by the normal acid.

*d.* In the examination of **soda-ash** it must be re-
membered that the presence of sulphides renders
the results obtained in saturation inaccurate. An
evolution of sulphuretted hydrogen on the addition
of dilute sulphuric acid indicates the presence of
**sodic sulphide.** If to dilute sulphuric acid a
drop of a solution of potassic permanganate, or of
potassic chromate, be added, and the colour of the
solution discharged in the first case and rendered
green in the latter, the presence of **alkaline sulphites or hyposulphites** is indicated.

If in these cases the soda-ash under examination
be first ignited with potassic chlorate, the results
will be rendered more accurate, since sodic sulphate
is formed; but the conversion of the sodic hyposulphite into sulphate is always attended with the
evolution of some carbonic anhydride, since 1 molec.
of sodic carbonate is decomposed in the process.

*e.* The estimation of **alkaline carbonate** may also
be effected by determining the carbonic acid by
one of the methods described under that head,
provided that only **one** carbonate be present, and
that as a **normal** salt. Normal carbonates of the
fixed alkalies and alkaline earths can be prepared
by heating with ammonic carbonate.

*f.* The estimation of the caustic **alkalies,** together
with the **alkaline carbonates,** is effected by
determining the carbonic acid in one portion that
has been ignited, and in another that has been
treated with ammonic carbonate. The difference
corresponds to a quantity of alkaline carbonate
that must be calculated as caustic alkali.

Or again, the total alkalinity may be first determined, and then the alkaline carbonate removed by adding a solution of baric chloride—

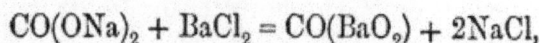

$$CO(ONa)_2 + BaCl_2 = CO(BaO_2) + 2NaCl,$$

and filtering off the precipitated baric carbonate. Wash the precipitate on the filter, and make up the filtrate and washings to a definite volume; take out a measured quantity, and determine the alkalinity in this with normal nitric acid; the presence of the baric chloride does not interfere with the accuracy. The alkalinity so found represents that due to caustic alkali. The carbonic acid in the precipitated baric carbonate may also be determined by means of normal nitric acid; this represents, of course, the alkalinity due to alkaline carbonate, and should coincide with the difference of the two previous determinations.

g. The estimation of the **alkaline earths** can also be effected alkalimetrically, in the same way as in the case of the alkalies. Normal nitric acid must be used; 1000 c.c. represent 76·5 grms. of baryta, 51·75 of strontia, 28 of lime, and 20 of magnesia.

The weighed quantity of the substance is dissolved in a measured volume of the normal acid; the acid-solution so obtained and coloured red with litmus is titrated with normal alkali until the blue colour appears.

In the case of carbonates of the alkaline earths, the substance must be dissolved in a flask in which the solution can be boiled so as to expel the carbonic anhydride ; the excess of acid is then, as above, titrated with normal alkali.

The estimation of the neutral carbonates can also be effected, as in the case of the alkalies, by the gravimetric or volumetric determination of the carbonic anhydride.

# IV. ELEMENTARY ORGANIC ANALYSIS.

UNDER this term is generally understood the determination of the elements **carbon, hydrogen,** and **nitrogen,** in organic compounds. As the processes, known as **combustions,** by which these elements are determined, are carried out with great uniformity, and irrespectively of the composition of the organic substance, it is possible to lay down very definite rules for the execution of all organic analyses in general.

In elementary organic analysis the organic substance is resolved into the simplest mineral compounds of its constituent elements, and then the amount of these compounds is determined by the ordinary processes of mineral analysis. Thus organic carbon is always converted into carbonic anhydride, hydrogen into water, and organic nitrogen into ammonia, or simply nitrogen gas. These products are then estimated either gravimetrically (always in the case of the water), or volumetrically, *i.e.* gasometrically; the latter method is particularly applicable when only minute quantities of the organic substance are available, as is the case in water analysis.

As no process for the direct determination of **oxygen** in organic compounds is known, it can only be estimated differentially after all the other elements have been determined.

## ORGANIC CARBON AND HYDROGEN.

These elements are usually determined in one and the same operation.

The oxygen, by means of which the carbon and hydrogen are, in combustion, respectively converted into carbonic anhydride and water, is supplied either **as such**, or as air, or is derived from some **oxy-compound** which readily parts with its oxygen at high temperatures, such as **cupric oxide, plumbic chromate**, etc.

*a.* **With cupric oxide only in a closed tube.**—A piece of combustion-tubing one-half inch in diam. and about 20 inches long is well washed internally, and afterwards thoroughly dried. One end is now drawn off obliquely to a point, known as a "sparrow-head," about 4 inches long, whilst the edge of the other extremity is rounded off by fusion in the blowpipe-flame. A small plug of ignited asbestos is now pushed down the tube to where it becomes narrowed; a little **dry** cupric oxide, coarsely powdered, and prepared as described below, is next introduced into the tube, so as to form a stratum between 1 and 2 inches in length. About ·2 grm. of the thoroughly-dried organic substance is weighed out into a porcelain dish or mortar containing about 10 to 20 grms. of finely-powdered dry cupric oxide, with which it is rapidly mixed by means of a pestle or glass rod. The mixture is now scooped into the combustion-tube, great care being taken that none is lost during the operation; the dish is then several times rinsed out with cupric oxide (at first fine and then coarse), which is also scooped into the tube, until the latter is filled to within about 2 inches of the open end. A second plug of ignited asbestos is pushed against the cupric oxide at the open extremity, and the aperture is closed with a cork until the combustion can be commenced.  It is essential that all the above operations should be performed with all possible despatch, to avoid the hygroscopic oxide of copper and the organic substance from absorbing atmospheric moisture.

The second portion of the apparatus consists of the **chloride-of-calcium-tube** and **potash-bulbs** respectively, in which the water and carbonic anhydride formed during the combustion are arrested and determined.

The **chloride-of-calcium-tube** is a U-tube (Fig. 4)

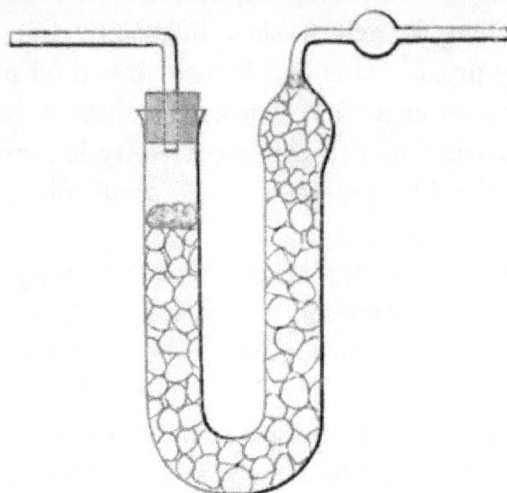

Fig. 4.

having a bulb blown on one limb, which is produced into a small horizontal tube, on which is blown another bulb; whilst the other extremity is fitted with an india-rubber cork bearing a small tube bent horizontally.  The limbs of the U-tube are filled either with pumice soaked in oil of vitriol

or with lumps of fused calcic chloride ; in the latter case a
stream of dry carbonic anhydride must be passed through

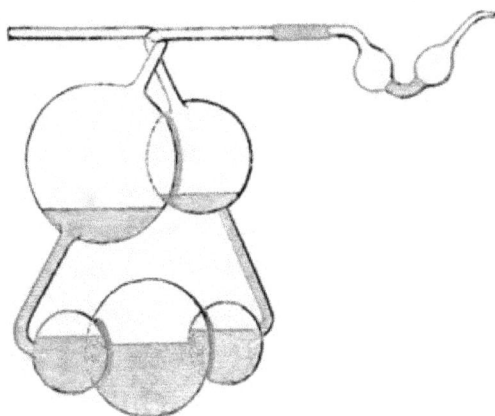

Fig. 5.

for one hour, in order that the calcic chloride may no longer
absorb any of this gas.   If oil of vitriol be used, great care

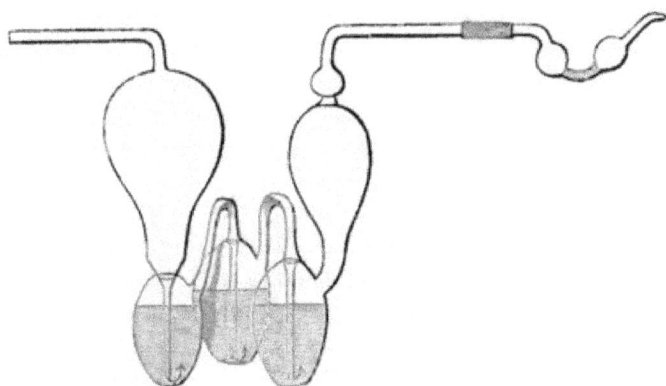

Fig. 6.

must be taken that it does not come in contact with the
india-rubber-cork.

The potash-bulbs. — These have either the familiar
form originally devised by Liebig (Fig. 5), or the modifi-

cation due to Geissler (Fig. 6), and which are purchasable from any chemical-apparatus dealer.   These bulbs are so blown that the requisite charge of the potash-solution is obtained by filling the large bulb at one extremity by suction.   The potash-solution to be used is prepared by dissolving caustic potash in its own weight of water.   To the exit-tube of the bulbs is attached a small tube with two little bulbs blown on it, and containing a few drops of strong sulphuric acid to arrest any moisture that might be carried over from the solution of caustic potash.

Both the potash-bulbs and the chloride of calcium tube should always be kept closed by means of glass stoppers, except when in use.

The combustion. — When the combustion-tube, the chloride-of-calcium-tube, and the potash-bulbs, have been charged as above, the two latter are carefully wiped outside with a dry cloth and accurately weighed.   The chloride-of-calcium-tube is then fitted by means of an india-rubber cork (that has been dried at 100° C.) to the combustion-tube, which is placed so that it projects about one or two inches from the combustion-furnace, and so that the "sparrow head" at the back stands upright.

The potash-bulbs are then attached to the exit-piece of the chloride-of-calcium-tube by means of a stout piece of caoutchouc tubing, which is then wired firmly with thin copper-wire ; the two glass-tubes should almost touch each other, leaving a minimum length connected by india-rubber only, as carbonic anhydride diffuses to a very considerable extent through caoutchouc.

The combustion-tube is now heated at the end nearest the cork, until the first few inches of the oxide of copper have acquired a dull red heat.   Heat is next applied to the small plug of oxide of copper behind the substance, and is then gradually continued from both ends of the tube to the part where the substance lies, until finally the whole length of the tube is at a dull red heat. The tube is maintained in this condition until no more gas is evolved; great attention must be bestowed throughout the operation upon the portion of the combustion-tube projecting from the furnace, as if this is too hot the cork will become charred, whilst if it is too cold some of the water formed will condense there instead of passing into the chloride of calcium tube. It may be laid down as a general rule that the projecting portion of the tube should be so hot that the finger cannot **permanently** be borne upon it.

When gas ceases to be evolved, a piece of india-rubber tubing is attached to the exit-tube of the potash-bulbs, and the furnace-burners near the "sparrow head" are turned out, the point of the "sparrow head" itself is broken off with a pair of pliers, and a gentle stream of air is drawn through the apparatus for several minutes by attaching the above piece of india-rubber tubing to an aspirator. By this means the water vapour and carbonic anhydride filling the combustion-tube are drawn over and absorbed in the chloride-of-calcium-tube and potash-bulbs respectively.

The chloride-of-calcium-tube and potash-bulbs are now detached and allowed to cool for twenty minutes in the balance-room, their apertures being kept closed by means of the stoppers already referred to. When cool they are again carefully wiped with a dry cloth and weighed. The increase in weight of the chloride-of-calcium-tube is due to the water formed by the combustion of the organic hydrogen; whilst the increase in weight of the potash-bulbs, and the small sulphuric-acid-bulbs which are weighed with them, is due to the carbonic anhydride derived from the carbon of the organic substance under examination.

If the organic substance contain nitrogen also, it is necessary that the combustion-tube should contain a roll of copper-gauze, about 4 inches long (see Analysis of Organic Bodies containing Nitrogen), which is placed between the cork and the coarse oxide of copper. This copper-gauze is heated up at first together with the front part of the oxide of copper, and serves to reduce any oxides of nitrogen that may be produced during the combustion, cupric oxide and free nitrogen being formed :—

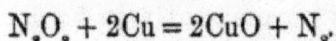

$$N_2O_2 + 2Cu = 2CuO + N_2.$$

**Preparation of oxide of copper for combustion.** —The purchased oxide should be heated to redness in a current of air in a glass or iron tube for about one hour. This may conveniently be done by having the tube open at both ends, and tilting the furnace at an angle of 15°. After ignition, the oxide should be kept tightly corked up until ready for use.

*b.* **With cupric oxide and oxygen in an open tube.**—The above method of combustion in a closed tube may generally be substituted with advantage by the following, in which the organic substance is burnt in a current of air or oxygen :—

The combustion-tube must be about 4 inches longer than the furnace, so that each end projects about 2 inches. A plug of asbestos is placed about 3 inches from the anterior extremity, or that to which the chloride-of-calcium-tube is attached, and coarse oxide of copper is then introduced at the other extremity, until a layer of closely packed oxide, about 18 inches in length from the asbestos, is obtained. A clear space of about 8 inches is left behind the oxide of copper, and then a "diffusion-tube," consisting of a piece of narrow hard glass tubing, sealed at one extremity, and about 3 inches in length, is placed with its sealed end forwards. A narrower glass-tube passing through an india-rubber cork fitted into the back opening of the combustion-tube penetrates into the open end of the "diffusion-tube."

When the combustion-tube has been arranged as above, it is gradually heated to dull redness throughout its whole extent, and a rapid current of air, from which the carbonic anhydride and moisture have been removed by means of caustic potash and oil of vitriol in a suitable drying apparatus, is passed through, entering by the tube mentioned above and leaving by the open anterior end of the combustion-tube.

After the tube has been heated to redness for about twenty minutes, the furnace-burners, excepting those under the oxide of copper, are turned out, and the posterior portion of the combustion-tube allowed to cool. In the meantime the dry organic substance is weighed out into a platinum or porcelain boat that has been recently ignited, and the chloride-of-calcium-tube and potash-bulbs are also weighed.

The chloride-of-calcium-tube and potash-bulbs are now attached to the anterior end of the combustion-tube, as already described in a; and the posterior end having by this time become cool (so that the hand can easily be borne upon it), the cork and "diffusion-tube" are removed, and the boat containing the organic substance is rapidly introduced and pushed forward until it is within a few inches of the oxide of copper. The "diffusion-tube" and cork are now replaced, the current of air reduced to about one bubble per second, and a couple of burners are lighted under the posterior end of the tube, well behind the boat, but somewhat in front of the "diffusion-tube."

The current of air heated by these burners at the back causes the organic substance to decompose; the volatile matter passing forwards into the red-hot oxide of copper is there converted into carbonic anhydride and water. The function of the "diffusion-tube" is to render the current of air behind the substance more rapid, and so to prevent its distilling into the posterior portion of the tube where it would escape combustion. Additional burners are from time to time lighted until the whole tube is heated to dull redness, and the current of air may be rendered more rapid towards the end of the combustion to burn off the carbonaceous mass in the boat; this, however, is more rapidly effected by exchanging the current of air for one of oxygen, also passed through the same drying apparatus.

When the carbonaceous residue is completely burnt off, the current of oxygen is again replaced by one of air, which is continued until the whole of the oxygen has been swept out of the tube; this can be ascertained by applying a glowing match to the exit-tube of the potash-bulbs; if the match does not ignite the oxygen is removed. The chloride-of-calcium-tube and potash-bulbs are now detached, and, after remaining for twenty minutes in the balance-room, weighed (see p. 81).

If the organic substance contain nitrogen, it is necessary, as in $a$, to place a copper-gauze cylinder in front of the cupric oxide to decompose any oxides of nitrogen that may be formed in the combustion. **Perkin** has replaced the

copper-gauze cylinder by a layer, about 6 inches in length,
of **pumice soaked in a solution of potassic chromate.**
I have found this just as efficient and much less troublesome than the copper, which, in open-tube combustions,
requires to be reduced in a current of hydrogen before each
operation.  The pumice should be broken into pieces
about the size of a pea, and then soaked in a saturated
solution of potassic chromate containing 10 °/$_o$ of dichromate.
After this the pumice is dried and ignited in a crucible,
and kept in a stoppered bottle ready for use.  During the
combustion it is only necessary that the layer of pumice
should be heated to a temperature short of redness.


## ORGANIC NITROGEN.

**Oxide of copper process—**

*a.* In the combustion with cupric oxide of all nitrogenous compounds, both organic and inorganic, the whole
of the nitrogen is liberated in the free state or as oxides
of nitrogen, which are reduced to free nitrogen by means
of the metallic copper cylinder placed in the front part of
the combustion-tube as described above.  Since the other
products of combustion—water and carbonic anhydride—
are condensible on the one hand, and removable by means
of caustic potash on the other, the nitrogen may be
determined by measuring the volume of gas, unabsorbed
by caustic potash, which is evolved during combustion.
This method, devised originally by **Dumas,** is, as stated
above, applicable to all nitrogenous bodies, both **organic**
and **inorganic.**  The process is carried out as follows :—

A piece of combustion-tube, about 40 inches long, is
sealed at one end.  A quantity of bicarbonate of soda, or

magnesite, dried at 100° C., is introduced into the tube so as to form a plug about 4 to 6 inches in length at the closed end. A small plug of asbestos is next introduced, and following this a layer of coarse oxide of copper about 2 inches in length. The nitrogenous substance (of which a quantity should be taken so as to yield about 10 c.c. of nitrogen—thus, ·1 grm. of a substance containing 10 °/$_o$ of nitrogen gives about 10 c.c of the gas) is then weighed into a glass or porcelain dish containing about fifty times its weight of finely-powdered and recently-ignited oxide of copper; with the latter it is now intimately mixed, and then the mixture scooped into the combustion-tube. The dish is then several times rinsed out with coarse oxide of copper, which is introduced into the combustion-tube in the same manner; this is continued until the tube is full within about 6 inches from its open extremity, a second plug of asbestos being then introduced, and finally the copper-gauze cylinder, which should have been recently ignited in a stream of hydrogen and allowed to cool in a current of dry carbonic anhydride.

The combustion-tube, when charged as above, is placed in the furnace so that about 2 inches of its open extremity project; it is then fitted with an india-rubber-cork and delivery-tube dipping into a little trough of mercury.

A single burner under the closed extremity of the tube is now lighted, upon which a rapid evolution of carbonic anhydride from the magnesite or bicarbonate of soda soon follows, and this, escaping by the delivery-tube, sweeps the air out before it. When this brisk current of gas has continued for fifteen minutes, a little should be collected in a small test-tube over the mercury; and if it is found that all but an exceedingly minute bubble dissolves in caustic potash, this shows that the whole of the air has been displaced. A measuring-tube, graduated into cubic centimetres and filled one-third with strong caustic potash solution (equal weights of potash and water) and two-thirds with mercury, is then placed over the delivery-tube to receive the gas evolved. The copper-gauze and the first few inches of oxide of copper are now heated to dull redness as rapidly as possible, whilst the heat under the closed extremity of the tube should be reduced so that only a bubble of gas escapes from time to time, and thus prevents the mercury from running back into the tube.

When the front part of the tube is red-hot, heat is applied to the oxide of copper placed behind the substance, and the burners between these two extreme points are now gradually lighted, one or two at a time, until the whole length of the tube, excepting that containing the carbonic-anhydride-generator, is at a dull red-heat. When this temperature has been maintained for about ten minutes, the closed end of the tube is again heated to obtain a current of carbonic anhydride to sweep out the nitrogen liberated. The current of carbonic anhydride is continued until the volume of fixed gas in the measuring-tube ceases to increase, and the bubbles of gas, as they enter the solution of potash, collapse with a sharp sound.

The gas in the measuring-tube is repeatedly shaken up with the potash, and after standing for about fifteen minutes the tube is removed by closing the end with a small crucible, and transferred to a tall glass cylinder filled with water; the crucible is now withdrawn, allowing the mercury to fall whilst the water rises to take its place. The tube is clamped in a vertical position, and so that the water is at the same level both inside and out. The gas is now at

the atmospheric pressure, and in this position it is left for
half-an-hour so that the water acquires the temperature of
the air. The volume is then read off as well as the atmospheric pressure and the temperature of the room. The
percentage of nitrogen may then be calculated as follows :—

$$G = V(b-w)\frac{0.0012566}{760(1+0.003665t)}$$

$$\frac{G \quad . \quad 100}{\text{Weight in grms. of substance employed.}} = \text{Nitrogen } \%.$$

G = Weight of nitrogen found.

V = Volume of gas.

w = Tension of aqueous vapour in mm. of mercury for
$t°$ C. (See p. 313).

b = Height of the barometer in mm.

t = Temperature in centigrade degrees of the water.

In the following table the values of $\frac{0.0012566}{760(1+0.003665t)}$

for $t = 0°$ C. to $t = 30°$ C. are given :—

| 0° | 0·00000165342 | 16° | 0·00000156183 |
|---|---|---|---|
| 1 | 164738 | 17 | 155644 |
| 2 | 164138 | 18 | 155109 |
| 3 | 163543 | 19 | 154578 |
| 4 | 162953 | 20 | 154050 |
| 5 | 162366 | 21 | 153525 |
| 6 | 161784 | 22 | 153005 |
| 7 | 161206 | 23 | 152488 |
| 8 | 160632 | 24 | 151974 |
| 9 | 160062 | 25 | 151464 |
| 10 | 159496 | 26 | 150957 |
| 11 | 158934 | 27 | 150453 |
| 12 | 158376 | 28 | 149953 |
| 13 | 157822 | 29 | 149457 |
| 14 | 157272 | 30 | 148963 |
| 15 | 156726 | ... | ... |

The above process gives results which are very fairly accurate, but a great improvement is effected by employing the **Sprengel pump** to remove the air and afterwards the nitrogen from the combustion-tube, the use of a carbonic anhydride generator, which is always objectionable, being thus avoided. The process of combustion *in vacuo* will, however, be fully described in the section on Water Analysis.

### Soda-lime process—

*b.* All nitrogenous substances, excepting the mineral and organic compounds of nitric and nitrous acids; the so called nitro-, nitroso-, azo-, and diazo- compounds of organic chemistry, give up the whole of their **nitrogen as ammonia** when heated with caustic alkali. This reaction constitutes the principle of the most convenient process for the determination of nitrogen in organic compounds, being the one which, whenever applicable, is most generally used. The process is carried out as follows :—

A piece of combustion-tube, about 20 inches in length, is sealed at one end, and then a layer, about 1 to 2 inches in length, of oxalic acid (dried at 100° C.) mixed with a little soda-lime is introduced into the closed end of the tube, and an equal bulk of freshly-ignited soda-lime is made to follow upon this. The substance is now weighed out into a porcelain dish containing sufficient soda-lime to fill about 10 inches of the tube, and with this it is intimately mixed. The mixture is then scooped into the combustion-tube, and the dish repeatedly rinsed out with fresh soda-lime, which is similarly introduced, the addition of soda-lime being continued until the tube is filled to within 3 inches of its open extremity, a plug of asbestos being then inserted. The combustion-tube is then tapped on the table so to as make a passage for the evolved gases. In the analysis of many substances, such as guanos, etc., which contain ammoniacal as well as organic nitrogen, it is better to introduce the substance into the tube direct, and then mix it there with the soda-lime by means of a stout iron or copper wire, the loss of ammonia being thus avoided.

A bulbed U-tube, fitted with india-rubber-cork and bent tube, or **Will and Varentrapp's** nitrogen-bulbs (Fig. 7),

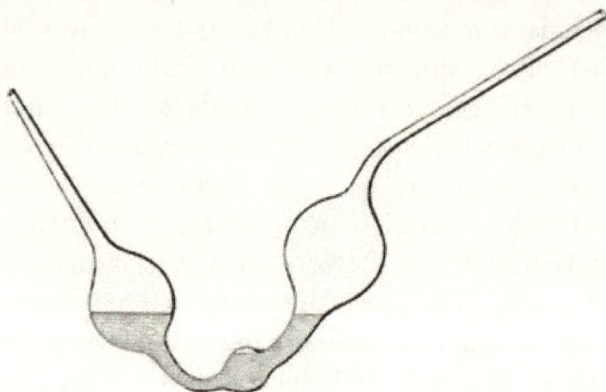

Fig. 7.

containing a known volume of standard sulphuric acid, is now fitted to the combustion-tube by means of an india-rubber-cork, and the combustion-tube placed in the furnace so that about 2 inches of its open extremity project.

The front part of the tube is now heated as rapidly as possible to dull redness; the soda-lime behind the substance is then similarly heated; the other burners, between these two extreme points, are then lighted, one or two at a time, until the whole length of the tube, excepting that with the oxalic acid at the end, is at a dull red-heat. The gas evolved bubbles through the sulphuric acid by which the ammonia is absorbed. When the evolution of gas has ceased, heat is applied underneath the oxalic acid, which, decomposing into carbonic anhydride, carbonic oxide, and water, drives out the ammonia which still remains in the tube.

The U-tube or bulbs containing the sulphuric acid are now detached, and the contents carefully washed out into the beaker into which the acid was originally measured ; a few drops of litmus-solution are now added, and then a standard solution of caustic soda is run in from a burette until the liquid is just rendered blue. The ammonia, and from it the nitrogen, can then be calculated from the amount of sulphuric acid neutralised during the combustion. If the substance contain ammonia, this must be deducted in order to obtain the organic nitrogen.

The nitrogen may also be determined **gravimetrically**. For this purpose the bulbs are charged with strong hydrochloric acid diluted with its own volume of water, and after the combustion this is evaporated to dryness on the water-bath, with an excess of platinic chloride. The dry mass is extracted with alcohol, and the insoluble double-salt washed on to a filter, which is then dried, ignited to bright redness in a tared porcelain crucible, which is then weighed with the residual spongy platinum. 194·4 parts of platinum are equivalent to 28 of nitrogen. The nitrogen cannot be calculated from the weight of the double-salt of platinum, as the latter generally contains similar compounds of platinum with organic bases ; since these salts, however, contain the same proportion of nitrogen to platinum as ammonic platinic chloride does, they do not interfere with the determination of nitrogen when the latter is calculated from the platinum left on ignition.

# V. EXAMINATION OF SOILS.

## 1. ARABLE SOIL.

AMONG the numerous factors which determine the fertility of arable land, two of the most important are certainly the chemical and physical nature of the soil. The chemical examination of a soil is intended to inform us what ingredients, both mineral and organic, are present, and in what quantity they are respectively found in the soil. The physical examination, on the other hand, deals with the state of aggregation in which the various constituents occur, together with their behaviour towards water, etc. A thorough chemico-physical examination of a soil should supply the agriculturalist with certain important facts which will enable him to treat the soil in the most advantageous manner. Thus—

1. The quantity of available plant-food, together with the composition of the reserve which will become available hereafter.
2. Determination of the constituents of plant-food which are absent in the soil, and the nature of the manures, etc., which can be most advantageously supplied.
3. Causes of the future sterility, either general or in respect to certain crops, of the soil.

In judging of the character of a soil it is necessary to take into consideration—

    *a.* Its geological origin.

    *b.* The nature of the subsoil from 4 to 6 feet.

    *c.* The climate and altitude above the sea-level.

    *d.* The position of the field.

    *e.* The previous treatment, rotation of crops, and manuring.

    *f.* Its previous fertility.

## I. PREPARATION OF THE SAMPLE OF SOIL.

In order to obtain comparable results on analysis, it is absolutely indispensable to use the greatest care in preparing a sample of the soil for examination. Owing to the great differences which soils exhibit in their upper and lower layers, it is impossible to lay down a perfectly general rule for the depth of soil which is to be included in the sample; moreover, the depth will vary according to the constituents which are to be determined; thus, in sampling a soil for nitrates it is advisable to include a greater depth than for the other ingredients, since the nitrates are readily washed down by rain. The depth should, however, always include that touched by the plough.

The following are the **directions issued by the Royal Agricultural Society for the collection of samples of soil for analysis** :—

"Have a wooden box made 6 inches long and wide, and from 9 to 12 inches deep, according to the depth of the soil and subsoil of the field. Mark out in the field a space of about 12 inches square; dig round, in a slanting direction, a trench, so as to leave undisturbed a block of soil with its subsoil from 9 to 12 inches deep; trim this block or plan of the field so as to make it fit into the wooden box; invert the open box over it, press down firmly, then pass a spade under the box and lift it up, gently turn over the box and nail on the lid. The soil will then be received in the exact position in which it is found in the field.

"In the case of very light, sandy, and porous soils, the wooden box may be at once inverted over the soil and forced down by pressure, and then dug out."

In the case of fields of uneven character, several samples should be taken from different parts and examined separately. A mixture of such samples can only by chance yield a correct average, and is only permissible in the case of fields of very uniform quality.

In order to render the sample of soil fit for examination, it must be spread out in a place sheltered from dust, etc., to dry. In damp and cold weather this process of dessication may with advantage be assisted by heating in an oven to 30 or 50° C.

## II. Mechanical Analysis.

### Relative Proportions of Gravel, Sand, and Clay.

1. Weigh the whole quantity or a representative fraction of the air-dried soil, rub it between the hands, and pick out any stones which may be present and weigh them, also note their size and mineralogical character.

2. Take 100 grms. of the soil and pass it through a sieve (No. 1) of copper-wire guaze, in which the meshes are about $\frac{1}{16}$ inch apart, powdering any earthy lumps

that refuse to pass in a mortar with a wooden pestle, and throwing the powder upon the sieve. Wash the residue with water until free from earthy particles, transfer to a dish, dry at 100° C., and weigh. This gives the **coarse gravel.** The mineralogical character of the residue should be noted. Ignite and weigh again; the loss in weight gives the quantity of organic matter in the coarse gravel.

The soil which has passed through sieve No. 1 is now transferred to a second (No. 2), the meshes of which are $\frac{1}{50}$ inch apart; here it is treated just as before, the residue being washed, dried at 100° C., weighed, ignited, and then weighed again. The amount of **gravelly sand** and organic matter mixed with it are thus ascertained.

Dry a portion of the soil which has passed through sieve No. 2 at 100° C., and take 15 grms. of it, boil it with water for about twenty minutes in a deep basin or flask. The boiling should be continued until all the particles have become completely separated from each other. The **coarse sand, fine sand,** and **clay,** are then separated from each other as follows :—Allow the boiled soil to cool, and then wash it into an **elutriating glass** (a tall champagne glass may be conveniently used) 7 or 8 inches deep, and about $2\frac{3}{4}$ inches wide at the mouth; a siphon-tube, just dipping but slightly into the glass proceeds from its side. A continuous but gentle stream of water is made to pass into the elutriating glass in such a manner as to cause a constant agitation of the particles, whereby the finest are washed away through the siphon-tube at the top of the glass, and received in a beaker, or other convenient vessel. This stream of water is best kept up and regulated by causing it to flow from a reservoir provided with a stopcock delivering into a funnel-tube, 12 to 18 inches long, drawn out to a fine aperture below.   The end of this funnel-tube is placed nearly at the bottom of the elutriating glass, and the supply of water to it is so regulated by the stopcock that the funnel-tube always·remains half-full of water.   When the water runs off through the siphon-tube nearly clear, the stopcock of the reservoir is closed, and the elutriating glass being removed, the water is decanted from it and the residue washed into a small dish, where it is dried and weighed, after which it is ignited and weighed again ; the two weights give the proportion· of coarse sand and its organic matter.   The elutriated turbid fluid is allowed to stand for several hours until it has become nearly clear. The supernatant liquid is poured off into another beaker, and the deposited matter, consisting of fine sand and fine soil, is then subjected to a second elutriating process, conducted as before, except that the force and volume of the washing water is considerably lessened.   The operation is continued until the wash-water passes off quite clear ; this sometimes takes three or four hours, but with the arrangements described above it requires no attention.   The residue in the elutriating glass is fine sand, which, with its organic matter, is estimated as before by drying, weighing, igniting, and re-weighing.   We have only now to deduct from the 15 grms. the quantities of coarse and fine sand, to obtain the proportion of finely-divided matter or clay.   The results of this mechanical analysis may be tabulated thus :—

100 parts of the soil, dried at 100° C., contain, for example,—

| | | | Fixed Substances. | Combustible or Volatile Substances. |
|---|---|---|---|---|
| 6·90 | { | Gravel (coarse) . . . | 6·90 | ... |
| | | Organic matter . . . | ... | 0·00 |
| 7·10 | { | Gravel (fine) . . . . | 6·43 | ... |
| | | Organic matter . . . | ... | 0·67 |
| 35·50 | { | Sand (coarse). . . . | 34·37 | ... |
| | | Organic matter . . . | ... | 1·13 |
| 40·00 | { | Sand (fine) . . . . | 38·50 | ... |
| | | Organic matter . . . | ... | 1·50 |
| 10·50 | { | Fine soil or clay . . | 9·50 | ... |
| | | Organic matter, ammonia, and combined water | ... | 1·00 |
| 100·00 | | | 95·70 | 4·30 |

Stones . . . . . 2·10 %.[1]

The same process is equally applicable to the mechanical analysis of clays, but in this case there is no necessity to look for gravel.

## III. Determination of the Physical Properties.

### 1. Specific gravity—

a. The weight of an empty "specific-gravity bottle" of 50 c.c. capacity is taken; it is then filled with distilled water at 16° C., and the weight again taken. Empty the bottle and introduce a certain quantity (10 to 20 grms.) of the soil (from which the big stones have been picked out) dried till constant at 100° C., and subsequent to weighing boiled with a little water to expel the air, fill the

[1] Sutton, *Volumetr. Analys.*

Sp.-G. bottle with water (16° C.), and weigh again.
The difference in weight between—

    *a.* The bottle filled with water together with the
       weight of the soil, and

    *β.* The bottle filled with soil and water, is equal
       to the weight of water displaced by the soil.
       The ratio of the weight of the soil to the
       weight of an equal volume of water is the
       specific gravity of the soil.   Thus, for
       example :—

Weight of Sp.-G. bottle + water = 140 grms.
        ,,     soil   .  .    = 20 ,,

                   160 grms.

Weight of Sp.-G. bottle + soil + water = 152 ,,

                    8 grms.

$$\therefore \text{ Sp. G.} = \frac{20}{8} = 2\cdot5.$$

    *b.* The specific gravity may also be approximately
    ascertained by weighing out 200 grms. of the soil
    into a ½ litre-flask, and then adding water from a
    100 c.c.-flask (filled to the mark) until the ½ litre-
    flask is full to the mark ; the water remaining in
    the 100 c.c.-flask is then measured with a pipette ;
    this represents the quantity of water, in c.c. or in
    grms., displaced by the given weight of the soil.

  2. Absolute weight—

An imperial half-pint is exactly filled with the dry soil
and weighed.  This weight multiplied by 150 gives very
nearly the weight of a cubic foot of the dry soil.

Soils rich in humus are light, whilst those containing much clay are very heavy.

The "**apparent Sp. G.**" of the soil is the ratio of the weight of a given volume of soil to the weight of the same volume of water. Thus, if 100 c.c. of the soil weigh 140 grms., the apparent specific gravity is 1·4.

**3. Porosity** is ascertained by calculating the percentage weight of dry soil contained in a given volume; this is done by dividing the apparent specific gravity by the real specific gravity, and calculating for 100 parts. Thus, in the above example:—

$$2\cdot50 : 1\cdot400 :: 100 : x = 56 ;$$

44 °/$_o$ of the volume of the soil is thus occupied by air.

**4. Capacity for water—**

Put two filters into the same funnel, then place 50 grms. of the air-dried soil upon the upper filter, and add cold water, drop by drop, to the soil until it begins to trickle down the neck of the funnel. Cover the funnel with a glass plate, and continue to add a few drops of water from time to time, until it is certain that the soil is perfectly soaked. Remove the filters from the funnel, and open them upon a linen cloth to remove the drops of water adhering to the paper. The outside filter is now placed in one pan of the balance, and the inner filter with the soil in the other, weights are added to the former pan until the balance is adjusted. The weight of the wet soil is thus obtained.

Thus, if 50 grms. of dry soil weigh 75 grms. when saturated with water as above, the soil is capable of holding 50 per cent of water.[1]

[1] Sutton, *Volumetr. Analys.*

### 5. Evaporation of moisture—

The moist soil on the filter (see 4, above) is exposed to the air upon a plate, and weighed at intervals of four, twelve, or twenty-four hours. In estimating the tendency of the soil to become dry from the rapidity with which it loses weight, account must be taken of the temperature and humidity of the air.

### 6. Permeability by water—

In a square zinc box, 12 inches high and 1 inch wide, with a funnel-shaped extremity and delivery-tube below, plug the latter loosely with cotton-wool, and fill the funnel-shaped portion with coarse sand. Moisten the sand and cotton-wool with water, and weigh the apparatus. Now fill the box about 8 inches deep with air-dried soil, and weigh again. Saturate the soil gradually with water, determine by weighing the quantity of water taken up, and then very carefully pour on 4 inches depth of water on to the wet soil, and note the time which it takes to run off. As in repeating the operation the time taken is always rather longer, several experiments should be made, and the mean of the results taken.

### 7. Capillary attraction for water—

By this is understood the length of time required for moisture to rise in the soil to a certain height (50 c.m.), or the height to which it rises in a certain time (48 hours).

Glass-tubes, about 80 c.m. long and $1\frac{1}{2}$ to 2 c.m. in diameter, and divided into centimetres along their length, are used for the purpose; at the lower extremity they are closed with a piece of fine linen, and the tube is then filled with the finely-powdered soil. The tubes are then placed vertically, with their lower extremities dipping a few milli-

metres below the surface of the water in a vessel, the level
of which is kept constant by the addition of water from
time to time.

It appears from experiments made by Krocker (*Schlesische*
*landwirthschaftliche Zeitung v. Janke*, Breslau, 1860, No.
2), in which graduated tubes 4 feet in length connected
with a reservoir of water below were used, and in which
the observations were continued over a period of from
twelve to sixteen months, that in some soils water is able
to rise even to such altitudes. Near the surface the
moisture did not appear until after about a year. The
quantity of water contained in the soil diminished with the
height. The tubes were afterwards cut up into lengths of
6 inches each, and the quantity of water determined in
each section. The following were the percentage quantities
of water thus contained in three different kinds of soil :—

| Height. | Clay Soil (after 16 Months). | Clayey Loam. (after 14 Months). | Loamy Sand (after 12 Months). |
|---|---|---|---|
| 4    feet. | 2·38 (dry). | 1·62 (dry). | 7·95 (moist). |
| 3·5  ,, | 2·38  ,, | 6· 0 (moist). | 9· 0  ,, |
| 3    ,, | 5· 0 (moist). | 11·10  ,, | 10·48  ,, |
| 2·5  ,, | 10· 0  ,, | 13·20  ,, | 11·10  ,, |
| 2    ,, | 17· 2  ,, | 15· 8  ,, | 11·52  ,, |
| 1·5  ,, | 18· 5  ,, | 18· 2  ,, | 12·30  ,, |
| 1    ,, | 19· 5  ,, | 20· 1  ,, | 13·04  ,, |
| 0·5  ,, | 23· 0  ,, | 22· 4  ,, | 16·10  ,, |
| 0    ,, | 27· 0  ,, | 26· 5  ,, | 19·50  ,, |

The same apparatus may also be used to ascertain to
what depth and in what time a given column of water
(5 to 10 c.m.) penetrates into the air-dried soil.

### 8. Absorptive power—

The power which the soil possesses of absorbing water from an atmosphere saturated with moisture is determined by spreading out 10 grms. of the air-dried soil evenly over the surface of a shallow zinc box placed above a vessel containing some water, the whole being enclosed, air-tight, by a bell-jar. On standing for several days, the temperature of the air being observed, it is again weighed; the increase in weight (calculated upon the soil dried at 100° C.) indicates the absorptive power.

### 9. Absorption and conduction of heat—

Fill a cubical zinc box (side about 3 inches) with very finely powdered soil ; surround the sides of the box with pasteboard, and place the whole in a wooden box open at the top ; expose the surface to the direct sunshine for some hours, and observe the temperature in the top centimetre layer.

To ascertain the depth and degree to which the absorbed heat penetrates, larger quantities of soil must be operated upon in the same manner.

To determine the conductive power of the soil, the latter is heated in the above vessel until a thermometer placed at the centre shows a temperature of 80° C.; the time required for the soil to cool to the temperature of the surrounding air (or 20° C.) is then observed.

## IV. Chemical Analysis.

Owing to the difference in the quality of soils being less due to a fundamental difference in their constituents than to a difference in the proportions in which these constituents

are present, the mere **qualitative** chemical analysis of a
soil is of but little value, inasmuch as almost all the same
bases and acids will be found in soils of the most opposite
agricultural character.  When such a qualitative examination is made, it is conducted according to the ordinary
schemes of qualitative analysis, which require no further
comment here.  On the other hand, the determination by
a **quantitative** chemical analysis of the proportions in
which the ingredients are present in the soil is of the
greatest value, more especially if further information be
gained as to the quantity of these ingredients which is
present in a form fit for immediate assimilation by plants,
and as to the quantity which is still in reserve, but which
will gradually become available for plant-nutrition in the
future.  Thus, if the analysis shows how much phosphoric
acid, lime, potash, magnesia, and nitrogen are immediately
available for plant-life, as well as the reserve store of these
ingredients which is present in the soil, a satisfactory opinion
can be formed by the agriculturist as to the manures which
should be applied in order that the fertility of the soil may
be improved.

According to the object in view it will be necessary to
determine more or fewer of the ingredients, and the character of the solvents will be varied.  Thus, in some cases
it may be necessary to determine the constituents soluble
in water, or in hot or cold hydrochloric acid, etc.  However the solvents may be varied, it is quite essential, in
order to obtain **comparable** results, that the same reagent
—and this in the same degree of concentration—should
be used.  Hence the importance in stating the results of
Soil Analyses to record particulars concerning the solvent
employed.

For general parposes the solvent to be used is boiling
hydrochloric acid of Sp. G. 1·15, and the chief ingredients
to be determined are potash, lime, magnesia, the ferric and
aluminic oxides soluble in hydrochloric acid, the total phosphoric acid, etc.   For the analysis, the finely-powdered air-
dried soil is employed.

### 1. Moisture—

About 5 grms. of the air-dried soil are placed between
watch-glasses, and dried in a cupboard at 100° C., with
the watch-glasses apart until the weight is constant.   The
loss represents the moisture.

### 2. Organic matter—

About 2 grms. of the residue obtained in (1) are placed
in a platinum crucible and ignited with the lid off, until the
whole of the organic matter has been destroyed.   When
the mass is cool it should be treated with a few drops of
pure ammonic carbonate, and heated to about 150° C. in
an air-bath.   By this means any carbonates that may have
been decomposed by the ignition are reconverted.   The
loss in weight represents the **organic** and other volatile
matter.

### 3. Silica—

Take 5 grms. of the air-dried soil and boil it in a
little flask for half-an-hour with pure hydrochloric acid,
Sp. G. 1·15.   Filter off the solution, wash the residue well
with boiling water, and reserve the residue for further
examination (**portion insoluble in hydrochloric acid**).
Evaporate the solution to dryness in a porcelain dish,
adding a little nitric acid to destroy the organic matter
and oxidise the iron.   Ignite gently until no more acid
vapours are given off.   Extract the mass repeatedly with

hot hydrochloric acid, filter off the silicic anhydride, wash, dry, ignite, and weigh as $SiO_2$.

### 4. Ferric oxide, alumina, manganese, lime, and magnesia—

Make up the filtrate in (3) to 500 c.c.

Take 200 c.c. of this solution (representing 2 grms. of air-dried soil) and precipitate the ferric oxide, alumina, and phosphoric acid, according to p. 41, 2, *b*, from a dilute acetic-acid-solution ; filter off ; wash the precipitate and dissolve it in hydrochloric acid ; divide the solution so obtained into two parts—in the one estimate the iron by means of decinormal bichromate-solution, and in the other precipitate with ammonia, and dry and ignite the precipitate obtained, after deducting the weight of ferric oxide. The alumina and phosphoric acid are calculated by difference. As it is generally unnecessary to differentiate between the iron and alumina, they may be estimated together instead of as above.

The filtrate from the acetic-acid-solution above, contains the **manganese, lime, and magnesia,** which are estimated as before described, the **manganese** being first precipitated with bromine, the **lime** in the filtrate by means of ammonic oxalate, and the **magnesia** in the filtrate from the lime with hydric disodic phosphate. (See p. 42, 4, *a.*)

### 5. Sulphuric acid and alkalies—

The remaining 300 c.c. of the half litre to which the filtrate from (3) was made up are used for these determinations. They represent 3 grms. of the air-dried soil.

As the solution from which the sulphuric acid is precipitated must contain but little hydrochloric acid, the greater part of the latter should be removed by evaporation. Dilute with water, heat to boiling, and keep the solution hot for some time after the addition of the baric chloride. Filter off the baric sulphate, and treat it according to p. 48. Treat the filtrate with ammonia, and use the precipitate, if necessary, for a second determination of the phosphoric acid. To the filtrate from the ammonia precipitate add ammonic carbonate and ammonic oxalate, and filter off the precipitated baryta and lime. Evaporate the filtrate to dryness.

The residue, obtained on evaporation, in which the alkalies are to be determined, is then ignited in a platinum dish. A concentrated solution of **pure oxalic acid free from alkalies,**[1] or **ammonic oxalate**, is then added, the whole evaporated to dryness and ignited in a platinum dish; after repeating this treatment extract with hot water —the insoluble magnesia ($MgO$) remains, whilst the carbonates of the alkalies pass into solution. Filter, and the filtrate (which must neither show any turbidity with carbonate or oxalate of ammonium, and which, on being evaporated to dryness and ignited, must be perfectly soluble in water) is evaporated to dryness with hydrochloric acid, and ignited in a tared platinum dish. The increase in weight represents the **alkaline chlorides** ($KCl$, $NaCl$). Determine the **potassium** by means of platinic chloride, and estimate the **sodium** by difference.

### 6. Phosphoric acid—

10 grms. of the air-dried and finely-powdered soil are

---

[1] Oxalic acid, free from alkalies, is prepared by dissolving oxalic acid in hot hydrochloric acid (10 to 15 per cent strength). When cold wash the small crystals with water, and recrystallise from water.

covered with concentrated nitric acid in a small flask (if
much carbonate of lime be present the acid must be added
very cautiously to prevent loss by frothing), and heated
carefully on the sand-bath until the evolution of nitrous
fumes has ceased, which shows that all the organic matter
is decomposed. Evaporate to dryness in a porcelain dish,
and ignite until no more acid fumes are evolved. Heat
the residue repeatedly with small quantities of dilute nitric
acid (1 pt. $NO_2(HO)$, 1 pt. $OH_2$), breaking up the lumps
with a glass rod, and pass the solution through a filter into
a beaker. Neutralise a large portion of the free nitric
acid, heat to about 70° C., and then add a large excess
of ammonic-molybdate-solution, and proceed as on p. 44,
$a$, $\beta$.

A duplicate determination of phosphoric acid may also
be made with the ammonia precipitate in (5) ; this should
be dissolved in nitric acid, and then precipitated with ammonic molybdate, or, if but little iron is present, it may be
dissolved in hydrochloric acid, and precipitated at once with
magnesia-mixture in the presence of citric acid (see p. 44).

## 7. Examination of the portion insoluble in hydro-chloric acid—

The insoluble residue in (3), consisting chiefly of silicate
of alumina and sand, is dried, detached from the filter, and
transferred to a platinum crucible ; the filter is incinerated
upon platinum-wire and added to the residue, the whole
being then dried at 100° C. and weighed.

Divide the dried residue into three parts, $a$, $b$, $c$, for the
following determinations :—

> $a$. **Mineral matters insoluble in hydrochloric
> acid.** —Ignite in a platinum crucible and weigh.

b. **Silica soluble in alkalies.**—Boil with a concentrated solution of sodic carbonate, to which some caustic soda has been added. Filter, acidulate the filtrate with hydrochloric acid, evaporate to dryness, ignite, and proceed as on p. 12.

c. **Substances decomposed by sulphuric acid.**— Treat with concentrated sulphuric acid in a platinum dish, heat and slowly evaporate the excess of acid, stirring frequently, until a dry powdery mass remains behind. Moisten with hydrochloric acid, boil up repeatedly with water, filter, and determine the silica, iron, alumina, lime, magnesia, and alkalies in the filtrate as before described.

The undecomposed insoluble residue from the above is dried, and a portion of it boiled with a solution of sodic carbonate and caustic soda as in b, the silica being determined in the filtrate, and calculated as **silica insoluble in hydrochloric and sulphuric acids.** This silica and that found in the hydrochloric and sulphuric acid extracts, together with the alumina found in these extracts, represents approximately the **anhydrous clay** in the soil.

d. **Decomposition by means of hydrofluoric acid.**—Powder the other portion of the residue in c, as finely as possible, in an agate mortar until all the particles are washed away with water; collect the washings, dry, and ignite gently. Then treat the finely-divided particles so prepared with **hydrofluoric acid** in the manner described in p. 13 (3, $\gamma$), and determine the bases and silica.

## 8. The reaction of the soil—

Lay pieces of red and blue litmus-paper on the moderately moist soil, cover the latter with a bell-jar, and observe the changes in colour, if any, which take place. Or some of the soil may be treated in a funnel with a small quantity of water, and that which has run through tested with litmus-paper.

If an **acid reaction** be obtained which disappears on exposing the paper to the air for some time, the acidity is then due to carbonic acid only, whilst if permanent it is generally due to free **humic acid**.

Some soils show an acid reaction even after gentle ignition. In such cases the acidity generally results from the presence of certain **sulphates**, such as those of alumina and iron, which possess an acid reaction.

The reaction of the soil should therefore also be tested after the organic matter has been destroyed by careful ignition.

**9. Carbonic acid** is estimated upon 10 to 20 grms. of the air-dried soil by means of either Schrötter's or the absorption-apparatus described on pp. 60 and 62.

It is often important to determine the carbonic acid both in the air-dried and in the ignited soil (after treatment with ammonic carbonate), as the calcic humate is converted into carbonate by ignition, and thus an estimate of the lime so combined can be formed.

## 10. Organic carbon, humus, and chemically-combined water—

As the loss by ignition of the soil dried at 100° to 120° C. is not only due to organic matter, but also to chemically-combined water, it is often of interest to make a direct determination of the quantity of organic matter.

*a.* Organic carbon is determined either by ordinary elementary organic analysis (the carbonic acid as carbonates found in (9) being subtracted from the carbonic acid obtained on combustion), or by the following method of Ullgreen, by oxidation with **chromic acid.** This process, which, in the absence of a combustion-furnace, is very convenient for soil analysis, is conducted as follows :—

5 or 8 grms. of the air-dried soil, together with a little water (20 grms.), are placed in a small flask, and 30 c.c. of strong sulphuric acid are then added to decompose the carbonates. The flask is fitted with a cork perforated by two glass-tubes bent at right angles—one of these just passes through the cork, whilst the other nearly reaches to the bottom of the flask. After allowing the flask to stand for some time, the first tube is attached to an aspirator, whilst the second tube is put in connection with a tube filled with soda-lime to remove all the carbonic anhydride from the air drawn into the flask by the aspirator ; the flask is thus freed from carbonic acid.

8 grms. of coarsely-powdered potassic dichromate (or 5 grms. of chromic acid, about 20 parts of chromic acid being added for every one part of organic matter supposed to be present in the soil) are then introduced into the flask, which is then rapidly closed and connected by means of the shorter tube with a small wash-bottle containing concentrated sulphuric acid, which is further connected with a U-tube containing pumice soaked in strong sulphuric acid, and this again is connected with weighed potash-bulbs (Liebig's or Geissler's, see Figs. 5 and 6, p. 79), to the end of which is attached a small bulb with a few drops of sulphuric acid, which is weighed with the potash-bulbs. (A

U-tube with soda-lime may also be used instead of the potash-bulbs.)

After closing the longer tube passing through the cork of the flask by means of a clamp and piece of india-rubber tubing, the flask is heated—at first very gently, as long as there is an evolution of gas, and finally nearly to boiling for some time. On removing the lamp the end of the potash-bulbs is, as before, connected with an aspirator, whilst a soda-lime-tube is again attached to the tube through which the air enters the flask. A few litres of air are in this way drawn through the apparatus. The increase in weight of the potash-bulbs represents the carbonic anhydride resulting from the oxidation of the organic carbon.

b. **Calculation of the humus and the chemically-combined water.**—If it be assumed that the humous substances free from nitrogen contain on an average 58 $\%$ of carbon, the quantity of anhydrous **humus** may be calculated by multiplying the carbonic anhydride found in (a) by 0·471. The difference between the loss on ignition of the soil dried at 100° C., and the calculated humus, together with the total combined nitrogen in the soil, represents the quantity of **combined water** not given off at 100° C.

c. **Humous substances soluble in water and alkalies.**—Place 10 grms. of the air-dried soil together with 200 c.c. of caustic-potash-solution ($\frac{1}{2}$ $\%$ strength) in a flask; weigh the flask with its contents, then heat to boiling, and restore the original weight when cool by adding distilled water. Filter, and measure off 3 to 5 c.c. of the filtrate into a flask, dilute the solution to about 100 c.c., and estimate the organic matter by means of potassic **permanganate** in the manner described under "Water Analysis." The results obtained are only of approximate accuracy, inasmuch as the dissolved humous substances are of very variable composition.

**11. Nitrogen (as organic nitrogen and ammonia)—** About 10 grms. of the soil dried at 100° C. are heated with soda-lime in a combustion-tube, in the manner already described for soda-lime combustions. Rather longer tubes than usual should be employed for the purpose.

**12. Ammonia—**

a. About 50 grms. of air-dried soil are placed in a clock-glass supported over a basin containing a measured quantity of sulphuric acid, the whole being enclosed air-tight by a bell-jar standing on a plate with a little mercury in it. The soil is moistened with milk of lime or caustic-soda-solution, and the bell-jar rapidly inverted over it. After standing forty-eight hours the normal acid is removed and titrated with normal soda, and the ammonia calculated from the quantity of acid neutralised. The soil is then stirred up with a glass rod and placed over a fresh portion of acid in the same way for another forty-eight hours.

The results obtained by this method are not very accurate, inasmuch as the caustic lime, soda, or magnesia used act upon the organic matter in the soil with evolution of ammonia, although by conducting the process in the cold, as above, the error due to this cause is reduced to a minimum.

*b.* Much more trustworthy results are to be obtained by the following method of **Schlösing's** :—

Prepare a solution of hydrochloric acid (1 litre of acid to 4 litres of water) and determine the ammonia which the solution accidentally contains by the method described on p. 21 (*a*).

100 grms. of finely-powdered air-dried soil are introduced into a flask of 1 to 2 litres capacity (according to the quantity of lime in the soil); 50 c.c. of the above dilute hydrochloric acid are then added in the cold, and, after the evolution of carbonic anhydride has ceased, another 50 c.c.; the flask is well shaken, and the addition of acid continued until the liquid is distinctly acid. It is quite essential to add sufficient acid to convert the whole of the alkalies and alkaline earths into acid salts, as only then the affinity of the soil for ammonia is destroyed.

If the supernatant liquid remains distinctly acid after repeated shaking, it is diluted with distilled water free from ammonia (see Water Analysis) until there are 400 c.c. of solution present; thus, if 100 c.c. of acid have been used, then 300 c.c. of the distilled water should be added. The flask, with its contents, after being tared upon a balance, is allowed to stand until the liquid becomes clear. This is usually the case in six to twelve hours, as the potassic chloride formed assists in the subsidence of the clay. The clear liquid is then decanted by means of a siphon provided with a pinchcock to regulate the flow. By re-weighing the flask the quantity of fluid decanted is ascertained; for example, let it be 302 grms. The residue in the flask is now collected upon a tared filter, and after washing and drying, the latter is weighed, from which the quantity of insoluble residue—56 grms. for example—is obtained. The empty

I

flask is then dried and weighed ; the difference from the original weight represents the total amount of soil and liquid, which we will suppose is 510 grms.   Then—

Weight of soil and liquid   =   510 grms.

„       decanted liquid  =   302   „

„       undissolved soil =   56   „

The total weight of the liquid after the decomposition of the soil is therefore $510 - 56 = 454$ grms.   Moreover, as the weight of liquid decanted for the determination of ammonia is 302 grms., therefore the weight of ammonia found in this must be multiplied by $\dfrac{510 - 56}{302} = 1\cdot5033$ that it may represent the ammonia obtained from 100 grms. of soil.

The ammonia in the decanted fluid is determined by distillation with magnesia that has been ignited to remove every trace of ammonia, in the manner described on p. 21 (a).

### 13. Nitric acid—

As the quantity of **nitrates** present in soils is usually very small, it is only by employing the most delicate methods for the estimation of this acid that reliable and satisfactory results can be obtained.   The methods which most adapt themselves to the determination of small absolute quantities are usually the **gasometric** ones, and thus, in the case of the nitric acid in soils, it is a modification of the already described method of Schlösing's that is the most applicable, and the one which, in places where large numbers of nitric acid determinations are made, is almost exclusively used.   This will therefore be the only method described here, whilst the so called **Crum-Frankland** method, which is also used for the determination of small quantities of nitric acid, will be fully described under "Water Analysis," inasmuch as it is more applicable to the estimation of nitric acid in waters than in soils. For less accurate determinations adaptations of the methods already described may be used, but whenever possible Schlösing's method should be adopted.

Schlösing's method of determining nitric acid has been modified and perfected with especial reference to soil analysis by Warington, who has shown that the accuracy of the results obtained by this process is very great (*Chem. Soc. Journ.* 1882, 351).

The first step to be taken is to bring the soil as quickly as possible into a dry state. If this is not done, the quantity of nitric acid found may greatly exceed that existing in the original soil, as nitrification will be continually in progress while the soil remains damp. It is of importance that the drying be conducted under particular conditions, for if wet soil be dried at a high temperature (approaching 100° C.) there is a tendency to a loss of nitrates, whilst slow drying at low temperature admits, on the other hand, of their production. The soil should be broken up immediately it is received from the field, spread upon trays in layers about 1 inch in thickness, and the trays placed in a stove-room kept at about 55° C.; the drying is then usually complete in twenty-four hours. At this temperature nitrification by an organised ferment does not occur, and, therefore, little or no nitric acid can be produced during the operation. After drying, stones and roots are removed, and the soil is finely powdered and placed in bottles. Although the samples thus prepared are not absolutely dry, yet the small amount of water left is insufficient to permit of organic change.

**Preparation of the watery extract.** — The old methods of shaking up a large quantity (50 to 1000 grms.) of the soil with water are open to many objections, for the amount of soil required is large, the time occupied very considerable, and the extract obtained is weak and turbid.

The method employed by **Warington** is to extract the soil by percolation on a vacuum-filter. A funnel 4¾ inches wide is made by cutting off the top of a Winchester quart-bottle; at the bottom of this funnel a disc of copper-gauze is laid, and on this two discs of filter-paper, the upper slightly wider than the one beneath. The filter is first moistened, and the dry powdered soil is then placed upon it; 200 to 500 grms. are taken, according to the supposed richness of the soil in nitrates. If the soil is of loose texture, it is shaken firmly together, but with a clay soil consolidation is better avoided. The funnel is now connected by a caoutchouc-stopper and glass-tube with a strong flask, water is poured on the soil, and the flask is put in connection with a water-pump. The water descends through the soil, and is collected in the flask. When 100 c.c. have passed through, it may be concluded that all nitrates have been extracted.[1] The collection of this extract may take from ten minutes in the case of a loose surface soil, to forty-five minutes in the case of a subsoil. The extract should be nearly clear.

**Analysis of the soil extract.** — The watery extract so obtained is placed in a small glass or porcelain basin, and

[1] See also *Jour. Roy. Agric. Soc.* 1881, 329, in which Mr. Warington records that more than three-fourths of the chlorides and nitrates were extracted from 7 lbs. of soil by the first 50 c.c.; whilst in the first 150 c.c. of solution the whole of the chlorides and 98·8 % of the nitrates were found.

evaporated nearly to dryness on a water-bath. The extract from an arable soil yields, when concentrated, a very small quantity of pale-brown syrupy liquid; a pasture soil, being much richer in organic matter, yields a more considerable extract. The extract is usually acid to litmus, and, when concentrated, often strongly so.

The apparatus employed for the determination of nitric acid in this extract differs but slightly from that already described on p. 54. Fig. 8 shows the form of apparatus which I employ, and which is very similar to that used by Mr. Warington at Rothamsted. The carbonic-acid-generator is formed of two vessels. The lower one, B, consists of a bottle with a tubular in the side near the bottom, and contains the marble from which the gas is generated. The upper vessel, A, consists of a larger similarly-tubulated bottle, containing the hydrochloric acid required to act on the marble. The two vessels are connected by an india-rubber-tube passing from the side tubular of the upper vessel to that of the lower one; the acid from the upper vessel thus enters below the marble. Carbonic acid is generated and allowed to escape at pleasure by opening a stopcock attached to the side tubular of the lower vessel, thus allowing hydrochloric acid to descend and come in contact with the marble. The fragments of marble used should have been previously boiled with water. The lower reservoir is nearly filled with the boiled marble, whilst the hydrochloric acid used in the upper reservoir should also have been well boiled before use ; and before introducing it into the upper vessel it should have some cuprous chloride dissolved in it, and as soon as introduced it should be covered with a layer of oil to prevent the absorption of oxygen. As long as the acid remains of an olive-tint,

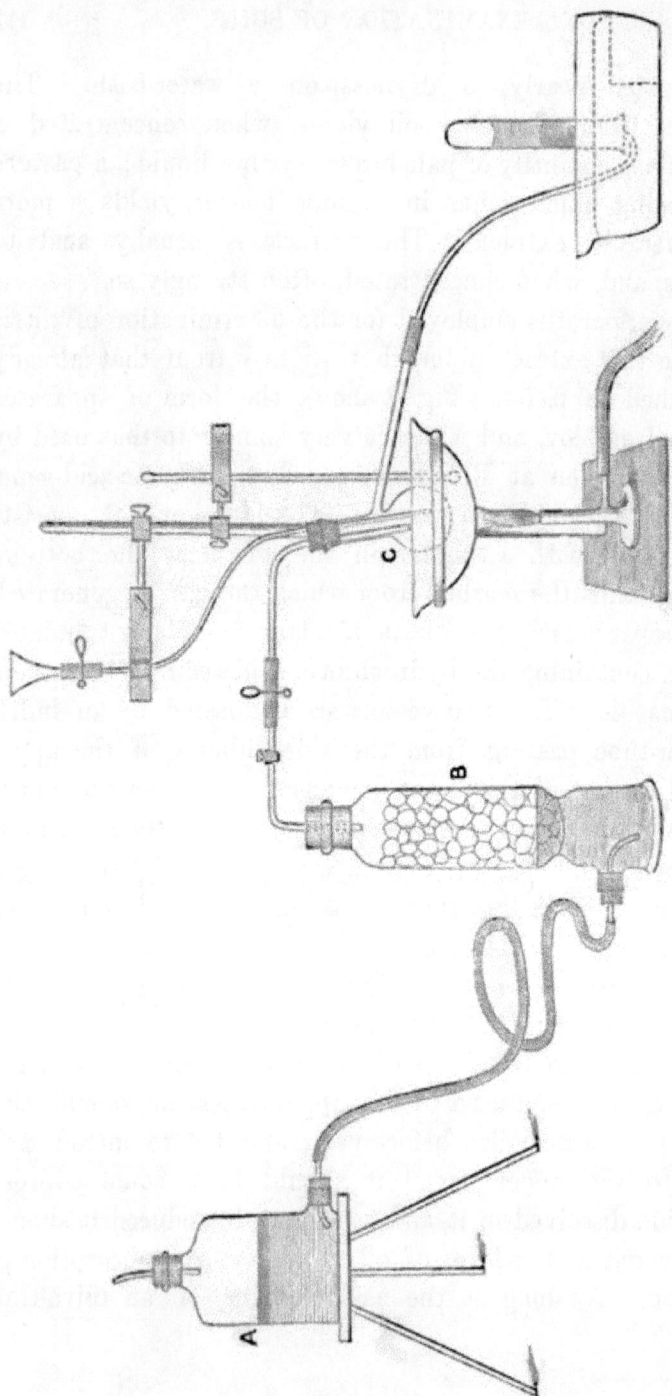

Fig. 8.

oxygen is absent, but if it change to a clear blue-green (due to cupric chloride) it is no longer certainly free from oxygen, and more cuprous chloride must be added. The reagents (hydrochloric acid and ferrous chloride) should also be boiled before use.

The apparatus figured opposite, p. 118, is fitted together, the long funnel-tube attached to the Wurtz-flask (which is only about 1¾ in. diam.) being filled with water. Connection is made with the glass-stopcock of the carbonic-acid-generator by means of a short stout caoutchouc-tube, provided with a pinchcock. The pinchcock being opened, the stopcock is turned till a moderate stream of bubbles rises in the mercury-trough; the stopcock is left in this position, and the admission of gas is afterwards controlled by the pinchcock, pressure on which allows a few bubbles to escape at a time. The Wurtz-flask is now almost submerged in an oil-bath, the temperature of which is kept at 130 to 140° C. By boiling small quantities of water or hydrochloric acid in the Wurtz-flask in a stream of carbonic acid the air present is soon expelled; the supply of carbonic acid must be stopped before the boiling has ceased, so as to leave little of this gas in the flask.

Soil extracts may be used without other preparation than concentration, Vegetable juices, which coagulate when heated, require to be boiled and filtered, or else evaporated to a thin syrup, treated with alcohol and filtered. A clear solution being thus obtained, it is concentrated over a water-bath to the smallest volume, in a beaker of the smallest size. As soon as cool, it is mixed with 1 or 2 c.c. of a cold saturated solution of ferrous chloride and 1 c.c. of hydrochloric acid, both reagents having been boiled and cooled immediately before use.

Great care must be taken that no bubbles of air become entangled in the reagents during mixing.

The mixture of the extract with ferrous chloride and hydrochloric acid is introduced through the funnel-tube, and rinsed in with three or four successive half cubic centimetres of hydrochloric acid. The contents of the flask are then boiled to dryness by means of the oil bath at 130 to 140° C., a little carbonic anhydride being from time to time admitted, and a more considerable quantity used at the end to expel any remaining nitric oxide. The gas is collected in a small jar over mercury. As soon as one operation is completed the jar is replaced by another full of mercury, and the apparatus is ready to receive a fresh extract.

The gas analysis is of a simple character ; the gas is measured after absorption of the carbonic anhydride by potash ; oxygen is then added, by which the nitric oxide is converted into nitrous anhydride and nitric peroxide ; both these are absorbed by the potash, and by adding a few drops of a saturated solution of pyrogallic acid to the latter the excess of oxygen is also absorbed. The residual gas is then measured, and the difference represents the volume of nitric oxide.

## V. Determination of Ingredients Prejudicial to Fertility.

### 14. Chlorine—

A measured quantity of the soil extract prepared for the estimation of nitric acid may be taken for the determination of chlorine. The solution is neutralised with sodic carbonate, evaporated to half its bulk, filtered, and the chlorine determined in the filtrate either by precipitation with dilute nitric acid and argentic nitrate, or volumetrically, with standard nitrate of silver and potassic chromate (see also 15).

The investigations of **Völcker** have shown that a soil becomes quite sterile if the quantity of common salt it contains rises above 0·1 °/₀ (*Journ. Roy. Agric. Soc.* 1865).

**15. Sulphur** (as sulphide and organically combined)— Certain clay soils, especially the blue clays of the Lias, often contain very considerable quantities of iron pyrites, rendering them sterile, until by oxidation the sulphur has become completely converted into sulphate.

In the course of analysis these compounds are not decomposed by dilute hydrochloric acid, but on ignition a portion of the sulphur is converted into sulphuric acid.

To determine the sulphur as sulphide together with the organically-combined sulphur, about 20 grms. of the air-dried soil are placed in a platinum dish and moistened with a saturated solution of potassic nitrate. After being carefully dried and ignited, the residue is extracted with hydrochloric acid and water. The solution, after filtration, is evaporated to dryness, and ignited to remove the silica; and on taking up again with dilute hydrochloric acid, the sulphuric acid is determined by precipitation with baric chloride.

From the sulphuric acid so found the quantity already discovered in the original hydrochloric acid extract of the soil must be deduced; the difference is due to the oxidation of the pyrites and "organic sulphur."

The same method of destroying the organic matter by ignition with nitre free from chlorides may be advantageously used in the estimation of **chlorine**, instead of as above.

50 grms. of the soil are gently ignited as above with potassic nitrate, and the residue, after cooling, is washed into a beaker, and the solution filtered. The filtrate is acidulated with acetic acid, and evaporated to dryness. Extract with water, filter, and determine the **hydrochloric acid** in the filtrate with argentic nitrate.

### 16. Ferrous oxide.—

Ferrous oxide, which results from the oxidation of the iron pyrites in the soil, may be estimated with approximate accuracy as follows :—

30 to 50 grms. of the finely-powdered soil are heated with twice the weight of hydrochloric acid (1·15 Sp. G.), together with a considerable quantity of ammonic chloride to prevent rapid oxidation. The operation should be conducted in a flask fitted with a cork perforated by a short glass-tube, to prevent the access of air. The contents of the flask are diluted with hot water, nearly neutralised with caustic soda, and the **ferric** oxide precipitated with sodic acetate (see p. 34).

The **ferrous** oxide, all of which is contained in the filtrate from the latter, is then oxidised with nitric acid, and the ferric oxide formed then precipitated as above and weighed ; from this the **ferrous** oxide to which it is equivalent is calculated.

### VI. STATEMENT OF THE RESULTS OF CHEMICAL ANALYSIS.

The results should be calculated upon the air-dried soil as well as upon the soil dried at 100° C. The solvents used should be specified, thus—

100 parts by weight of the finely-powdered soil dried at 100° C. contain—

1. Substances, combustible, decomposable, or volatile on ignition. (Loss on ignition.)

- *a.* **Humus**, calculated from the organic carbon—
  - *α.* Soluble in water.
  - *β.* Soluble in alkalies.
  - *γ.* Insoluble.
- *b.* **Nitrogen**—
  - *α.* As ammonia.
  - *β.* As nitric acid.
  - *γ.* Organic.
- *c.* **Water**, chemically combined.

2. Soluble in water and hydrochloric acid. (Sp. G. 1·15.)

- Alumina.
- Ferric oxide.
- Ferrous oxide.
- Manganous oxide.
- Calcic carbonate.
- Lime.
- Magnesia.
- Potash.
- Soda.
- Phosphoric acid.
- Sulphuric acid.
- Chlorine.
- Silica.

3. Residue from above soluble in alkalies　.　Silica.

4. Residue decomposable by concentrated sulphuric acid. (Clay.)

- Silica.
- Alumina.
- Lime, etc.

5. Residue decomposable by hydrofluoric $\left\{\begin{array}{l}\text{Silica,}\\ \text{Alumina,}\\ \text{etc.}\end{array}\right.$
    acid.   (Silicates, quartz.)

The following is the composition of a heavy clay pasture soil, recently analysed by me, from the county of Surrey :—

|  | Air-dry. | Dried at 100° C. |
|---|---|---|
| Water . . . . . | 6·39 | ... |
| Organic and volatile matter . | 8·26 | 8·82 |
| Matter insoluble in hydro-chloric acid . . | 64·30 | 68·58 |
| Silica . . . . . | ·05 | ·06 |
| Alumina and ferric oxide . . | 6·33 | 6·76 |
| Lime . . . . . | 8·58 | 9·17 |
| Magnesia . . . . . | ·31 | ·33 |
| Phosphoric acid . . . | ·05 | ·05 |
| Nitrogen . . . . . | ·15 | ·16 |
| Potash . . . . . | ·20 | ·21 |
| Soda . . . . . | ·11 | .12 |
| Carbonic acid, etc. . . . | 5·27 | 5·74 |
|  | 100·00 | 100·00 |

## 2. MARL, LIMESTONE, ETC.

The agricultural value of the substances included under the terms marl and limestone consists chiefly in the proportion of carbonate of lime they contain. The determination of this constituent is generally all that is required, but sometimes it may be necessary to make a complete analysis, which is then conducted upon much the same plan as that already described for soils.

Before commencing the analysis the air-dried sample must be **finely powdered** in an agate mortar.

### 1. Moisture—
About 3 grms. of the finely-powdered substances are dried at 100° C. The loss represents the moisture.

### 2. Carbonic acid
is determined in the same way as described in the Analysis of Soils, a much smaller quantity (about 1 to 2 grms.) of the substance being used.

A more rapid, and for many practical purposes sufficiently accurate, method may be substituted for the above. About 2 to 3 grms. of the substance are weighed into a small flask (about 100 c.c. capacity), and covered with a little water; a short test-tube half-filled with hydrochloric acid is then placed in the flask in an erect position, so that the acid remains in the tube. The whole is then weighed. The flask is now inclined so that some of the acid runs out, which, on coming in contact with the substance on the bottom, causes an evolution of carbonic anhydride; the remainder of the acid in the test-tube is gradually added in this way until the whole of the carbonate is decomposed. The flask is then again weighed; the loss represents the carbonic anhydride evolved.

### 3. Lime—
Assuming that the whole of the carbonic acid is combined with lime, the quantity of the latter may be calculated from the carbonic anhydride found, 44 parts of carbonic anhydride corresponding to 56 parts of lime.

The lime may also be approximately estimated by titration with normal nitric acid. (See Alkalimetry.)

**4. Substances** soluble in hydrochloric **acid.—** (Silica, lime, magnesia, alumina, iron, alkalies, sulphuric and phosphoric acids.)

These are determined according to the methods already described in the Analysis of Soils. A smaller quantity of substance (2 grms.) may with advantage be taken for the estimation of **lime, magnesia, iron, and alumina,** whilst a larger quantity (about 20 grms.) should be used for **phosphoric acid** and **alkalies.** The **sulphuric acid** should be estimated in a special portion (10 grms.) by boiling with a solution of pure sodic carbonate, filtering and precipitating the sulphuric acid in the filtrate with baric chloride after acidulating with hydrochloric acid.


5. **Clay and sand—**

The portion insoluble in hydrochloric acid is collected on a filter and washed until the filter is no longer acid ; the residue is then dried, ignited, and weighed; it is calculated as **clay and sand.**

A further discrimination between the clay and sand is made by treating 20 to 30 grms. of the marl, etc., with hydrochloric acid until all the carbonates are decomposed ; the liquid is then filtered, and the residue after being washed is boiled for half-an-hour with water, and then treated in an **elutriating glass** in the same way as described in the Mechanical Analysis of Soils.

In many cases a more rapid but less accurate method of estimating the clay and sand will suffice. 10 grms. of marl are treated with hydrochloric acid to completely decompose the carbonates ; the liquid is then stirred up with a large quantity of water, and, after allowing the sandy particles to subside, the supernatant liquid bearing the clay in suspension is poured off. This process of agitating the residue with water and decantating is repeated several times until the supernatant liquid is clear. The residue, consisting of the sand, is then collected on a filter, dried, ignited, and weighed.

The difference between the above weight of sand and the weight of the mixed clay and sand in (5) gives the weight of clay and fine sand.

### 6. Organic and volatile matter—

This is determined in the same way as in soils, by ignition in a platinum dish or crucible, and subsequent treatment of the residue with pure ammonic carbonate, and then heating from 130 to 150° C. to volatilise the excess of the latter. From the loss on ignition must be subtracted the moisture found in (1), p. 125.

### LIME.

The moisture, carbonic acid, and insoluble residue, are determined according to the methods already mentioned in the Analysis of Marl, Limestone, etc.

Estimation of the caustic lime.—The lime is finely powdered as rapidly as possible to prevent absorption of carbonic anhydride. 2 to 3 grms. of the powder are weighed into a flask of about 200 c.c. capacity. A moderately-concentrated solution of ammonic nitrate is then prepared in another flask, and the solution is boiled to expel any carbonic anhydride that may be held in solution. The hot solution of ammonic nitrate is now poured into the flask containing the lime and corked up. The flask is then shaken for a few minutes to accelerate the reaction.

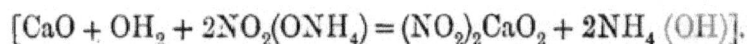

$$[CaO + OH_2 + 2NO_2(ONH_4) = (NO_2)_2CaO_2 + 2NH_4 (OH)].$$

The time required for its completion will depend upon the state of division of the lime. When the reaction is finished the residue should appear somewhat gelatinous; it is allowed to subside, and the supernatant liquid—an ammoniacal solution of calcic nitrate—is decanted through a filter; the residue is washed once or twice by decantation, and finally on the filter with boiled water. In the filtrate the lime is precipitated with **ammonic** oxalate, and further treated, as already described on p. 28.

The above operations must all be performed with promptitude, and air excluded from the solutions as much as possible.

## CLAY.

The several varieties of clay, all of which result from the decomposition of silicates, contain, besides aluminic silicate, other ingredients varying according to the origin of the clay, and upon which the technical value of the clay depends. According to their physical nature, the constituents of clay are divided into **plastic** and **non-plastic**, the latter being chiefly of a sandy character. Upon the chemical composition of this sandy part (whether quartzose or felspathic) depends the **fusibility** of the clay. Thus, as a general rule, those clays which contain a larger percentage of quartz are more **fire-proof** than those which are richer in undecomposed felspar; for the greater the proportion of bases, especially the alkalies and lime, and then magnesia, manganese, and iron, the more fusible the clay becomes.

MECHANICAL ANALYSIS.—The clay, after being boiled for some hours with water, is treated in an elutriating glass in the manner already described in the Mechanical

Analysis of Soil.    The sandy non-plastic portion thus separated should be further examined microscopically.

CHEMICAL ANALYSIS.    1. **Water—**

The air-dried clay contains both moisture, which can be expelled at 100° c., and water chemically combined, which is only given off at high temperatures.

The loss on ignition includes, besides the water, also organic matter, which is generally present in small quantity.

## 2. General chemical composition—

Heat 10 grms. of the clay with hydrochloric acid, filter, and dilute the filtrate to 500 c.c.    Examine one portion of this solution qualitatively.    Since the alkalies are determined after decomposing the whole of the clay, the present examination is limited to the determination only of lime, magnesia, iron, alumina, carbonic and sulphuric acids, which is done according to the methods given in the Analysis of Soils.

## 3. Decomposition with sulphuric acid, and determination of chemically-combined silica, amorphous silica (hydrated), and quartz or sand—

    *a.* Dry 5 grms. of the air-dried clay at 100° C. to determine the moisture, and then ignite for the estimation of combined water and organic matter. Take 2 grms. of this residue and moisten them with water in a platinum dish; now add four or five times the weight of strong sulphuric acid, and heat until the excess of acid is driven off. (The sulphuric acid in this operation displaces the silicic acid from its combination with bases, the silica so liberated being insoluble in water and dilute acids.) The evaporation with sulphuric acid may with advan-

tage be repeated a second time. The dry mass is
heated in the dish with water and a little hydrochloric acid; the insoluble residue is filtered off,
washed, dried, and ignited. (Sand and amorphous silica.) The filtrate serves for the determination of the bases.

b. The residue from a, consisting of sand and amorphous silica, is boiled with a concentrated solution
of sodic carbonate and filtered hot. The amorphous silica passes into solution; the undissolved
sand, after being washed with hot water acidulated
with a little hydrochloric acid, is dried, ignited, and
weighed; the loss in weight from a represents the
amorphous silica, the remainder the sand.

c. The hydrated silica contained in the clay is
determined by boiling a larger quantity of the clay
with a solution of sodic carbonate, filtering, and
separating the silica in the filtrate by evaporation
with hydrochloric acid.

4. Decomposition with fusion-mixture ($CO(ONa)_2$,
$CO(OK)_2$), and determination of the total silica, together with the bases, excepting the alkalies—

Fuse 1 or 2 grms. of the clay dried at 100° C. with
fusion-mixture, in the manner described on p. 12. After
treatment with hydrochloric acid, the silica is separated
by evaporation and ignition, and the bases are determined
in the extract of the residue, according to the methods
given in the Analysis of Soil.

5. Decomposition with baric hydrate or ammonic
fluoride for the determination of the alkalies—

1 or 2 grms. of the clay dried at 100° C. are decomposed

in the manner already described on p. 12. The hydrochloric-acid-solution of the bases so obtained is (when the
decomposition has been effected with ammonic fluoride, the
sulphuric acid must first be precipitated with baric chloride)
treated with ammonic carbonate and ammonia, and the
precipitate, after cooling, filtered off. The filtrate, after
being evaporated to dryness and ignited to drive off
ammonia-salts, is, after separating any lime or magnesia
that may be present, used for the determination of the
alkalies according to the ordinary method.

BEHAVIOUR AT A HIGH TEMPERATURE.—The fusibility
of clay may be roughly ascertained by means of the blowpipe. The moist clay is shaped into small cones, which,
after being thoroughly dried at the ordinary temperature
of the air, are exposed at their points to the blowpipe-flame.
The extent to which the sharpness of the point of the cone
is rounded off by the operation indicates the relative fusibility of the clay.

# VI. ANALYSIS OF PLANTS AND VEGETABLE STRUCTURES.

THE processes of Elementary Analysis already described would enable us to ascertain the elementary composition of plants and vegetable structures. Organic analysis would enable us to determine the proportions in which the elements carbon, hydrogen, and nitrogen are present in organic combination; whilst an ordinary mineral analysis would inform us of the presence or absence of any of the other elements—both metallic and metalloid. The elementary composition of plants possesses, however, generally but a subordinate interest, whilst of far greater importance for practical purposes is the determination of the states of combination in which these elements occur. Although this species of organic analysis is so far but comparatively undeveloped, yet many ready methods for the detection and estimation of some of the more important organic compounds met with in the vegetable kingdom are now at the disposal of the analyst. This section contains, firstly, the processes for determining the mineral ingredients of plants; secondly, those for the determination of some of the more common vegetable organic compounds; and thirdly, the application of these methods to the analysis of special products of agricultural interest.

# 1. THE ASHES OF PLANTS.

The analysis of the ashes of plants reveals the mineral substances which vegetable growth removes from the soil. Neglecting those substances which are of very exceptional occurrence, the chief ingredients of the ashes of plants are : —Potash, lime, magnesia, ferric oxide, silica, phosphoric acid, sulphuric acid, carbonic acid, and chlorine, besides carbon and sand, the latter being derived from dust adhering to the plants.   Thus alumina, which is such a universal constituent of soils, is rarely if ever met with in the ashes of plants.   The carbonic acid is generally only formed in the carbonisation of the plants, the salts of the organic acids yielding carbonates on ignition.   Thus the carbonic acid, as well as carbon and sand, found in the ash are not derived from the soil in which the plant grew.

In the preparation of the ash it is necessary—

(1) That the plants or vegetable structures be freed as much as possible from all adhering dust, etc.

(2) That the ash contain as little unburnt matter as possible, and that no part of the essential ingredients be lost during the incineration.

These conditions are secured by thorough cleaning on the one hand and careful ignition on the other.

Thus roots, tubers, etc., are cleansed by wiping with a wet linen cloth, whilst grain is washed with water in a basin, from which it is afterwards separated by a sieve and dried with a linen cloth.

The substance is then dried, roots and tubers being cut into thin slices and hung up in drying cupboards by threading them on to a piece of cotton, and finally placing

them in a beaker or dish.   The air-dried substance is then
reduced to a coarse powder; stalks, etc., being cut into
small pieces, and the whole well mixed.


## METHODS OF IGNITION.

1. The dried substance is heated in a platinum dish
placed in a muffle-furnace, kept at a temperature below
redness.

In the case of the ashes of trees, feeding plants, turnips,
etc., which are rich in carbonates, the ignition may be
conducted in a platinum dish over a lamp; the carbonisation is first carried on at a very low temperature, and
then the combustion of the carbonaceous mass is effected
by increasing the heat.

The crude ash is weighed; 1 grm. of it is employed
for the determination of **carbonic acid**.   The residue
insoluble in acids is collected on a filter, dried, and any
**carbon** that it may contain estimated by the loss on
ignition.

The quantity of carbon and carbonic acid found is
calculated on the whole of the ash and subtracted from the
latter, together with any **particles of sand** that may have
been discovered; in this way the weight of the **total ash**,
free from carbon, carbonic acid, and sand, is ascertained.

2. In the case of substances which, like the seeds of
plants, are very difficult to reduce to ash, it is advantageous
to simply carbonise in the first instance in a platinum dish
at a low temperature, then to powder the carbonaceous
mass in a mortar and extract with water; the insoluble
residue is now dried, and then completely burnt by stronger
ignition, which generally involves no difficulty.   The

ignited residue is weighed, the aqueous solution is added
to it, evaporated to dryness, ignited, and the whole weight
obtained represents the **total ash.**

3. The carbonaceous residue obtained on gentle ignition
can often be completely burnt off in the platinum dish by
placing above the latter a platinum triangle bearing a small
glass cylinder (a lamp-chimney or neck of a retort answers
the purpose well), by which means a sufficient draught
of air over the ignited mass is secured.

A portion of the crude ash thus obtained is treated as
in (1) to obtain the weight of the ash free from carbon and
carbonic acid.

4. **By addition of baric hydrate.**—In the case of
easily-fusible ashes rich in silica, a loss of sulphur and phosphoric acid may easily take place on ignition.   In order to
prevent this loss, the substance dried at 100° C. should be
carbonised at a low temperature, and the carbonaceous
mass treated with a concentrated solution of baric hydrate,
so that the ash contains about half its weight of baryta.
After drying, the whole is ignited in a muffle, and the
residue powdered and well mixed.

The ashes may be divided into the following groups,
according to their chief constituents :—

  a. Ashes containing principally the **carbonates of
     the alkalies and alkaline earths.**   (Wood and
     ordinary vegetables.)
  b. Ashes containing principally **phosphates of the
     alkalies and alkaline earths.**   (Seeds.)
  c. Ashes containing principally **silica.**   (Stems of
     grasses.)

I. Examination of Ashes containing principally
Compounds of Carbonic and Phosphoric Acids.

As the nature of the chief constituents will vary to a
certain extent the course of analysis adopted, it is necessary, in the first place, to make a few preliminary observations in order to ascertain to which of the above groups
the ash in question belongs.

The ashes of group *a* effervesce briskly on treatment
with hydrochloric acid.

The hydrochloric acid extract of an ash, after separation
of the silica, generally yields, on carefully neutralising with
ammonia and then adding acetic acid in excess, a yellowish-
white precipitate of ferric phosphate [$Fe_2(PO_4)_2$]. If the
filtrate from this gives with ammonia a further precipitation (phosphates of the alkaline earths), it shows that
there is more phosphoric acid than can be combined with
ferric oxide present ; this is generally the case in ashes
of group *b*.

The ashes of group *c* are often not completely soluble
in hydrochloric acid.

## 1. Determination of Silica, Sand, and Carbon.

Moisten 2 or 3 grms. of the ash in a porcelain dish with
nitric acid, and then add hydrochloric acid until effervescence ceases. (In order to prevent loss by spirting, the
dish should be covered with an inverted funnel, through
the tube of which the acid is added by means of a smaller
funnel.) Evaporate to dryness on the water-bath, and
extract the dry mass with water and a little hydrochloric
acid. **Sand, carbon, and silica** remain undissolved ;

this residue is collected upon a weighed filter, washed, dried at 110° C., and weighed.

The residue is then detached from the filter-paper and heated with a mixture of a solution of sodic carbonate and dilute caustic soda ; the silica dissolves.  The residue from this is collected on a weighed filter, washed, dried, and weighed.  The loss in weight represents the silica, which may be also determined directly by separation from the alkaline solution.  The remainder is carbon and sand. More than ·4 grm. of carbon must not be present, as an appreciable quantity of soluble salts may then be retained.

## 2. Determination of Chlorine, Carbonic Acid, and Sulphuric Acid.

The carbonic acid is determined in 1 grm. of the ash by decomposition with nitric acid, according to one of the methods described on p. 60.  The chlorine is estimated in the diluted filtrate from the above, either volumetrically or gravimetrically, by means of argentic nitrate.

As during the ordinary process of incineration there is an appreciable loss of chlorine, the above determination will not suffice in cases where great accuracy is requisite. In such cases 10 grms. of the dried substance should be moistened with a solution containing 1 grm. of pure sodic carbonate ; after drying, the whole is reduced to ash by ignition.  The ash is then dissolved in nitric acid, and the chlorine determined by argentic nitrate.

The chlorine, together with the small quantity of sulphuric acid ready formed in the plant, can be simultaneously determined by extracting the dried and finely-divided substance with water acidulated with nitric acid.  In the solution obtained the sulphuric acid is determined with baric nitrate, and the chlorine in the filtrate from this with argentic nitrate.

To ascertain the total sulphur, both as sulphuric acid and as organically-combined sulphur, 4 or 5 grms. of the finely-divided substance are fused with six times the weight of the purest caustic potash and 2 or 3 grms. of potassic nitrate. The latter are first fused in a silver crucible or dish, and the substance then added, in small quantities at a time, and oxidised. The mass, when cool, is dissolved in hydrochloric acid, and, after separation of the silica, the sulphuric acid is precipitated with baric chloride and weighed.

### 3. Determination of the remaining Ingredients.

Divide the solution into two parts, and determine in Part A.—Phosphoric acid, ferric oxide, manganese, lime, and magnesia; in Part B.—Alkalies and sulphuric acid :—

EXAMINATION OF PART A.—Treat the cold acid-solution with ammonia until just alkaline, and then add acetic acid in excess; a precipitate of ferric phosphate, $Fe_2(PO_4)_2$, is thrown down. Filter off, wash with water, dry, ignite, and weigh the precipitate. Precipitate the lime in the filtrate by means of ammonic oxalate, filter off, dry and ignite, weigh as carbonate or quicklime. The filtrate, if it still contains phosphoric acid, is divided into two portions (if it does not contain phosphoric acid, the magnesia is determined in the whole, according to 1), and—

  1. Portion A is rendered alkaline with ammonia, and then hydric disodic phosphate added; the magnesia

is precipitated as ammonic-magnesic phosphate, and
after collecting on a filter, washing with a mixture
of one part strong ammonia and three parts water,
drying, and igniting, it is weighed as magnesic
pyrophosphate.

2. Portion B is rendered alkaline with ammonia and
magnesia-mixture added; the phosphoric acid is
precipitated as ammonic-magnesic phosphate, which
is filtered off and further treated as in 1.

If larger quantities of iron are contained in the ash,
or if manganese be present, the following course should
be adopted :—

The hydrochloric-acid-solution is diluted with water,
and treated with sodic carbonate in the cold until the precipitate which forms is only just redissolved on shaking; a
strong solution of sodic acetate is then added, and the
liquid boiled for fifteen minutes. When the precipitate
has subsided the solution is filtered hot.

a. The precipitate, consisting of basic ferric phosphate and acetate, is dissolved in nitric acid,
and the solution so obtained divided into two parts,
in one of which the phosphoric acid is determined
by means of ammonic molybdate, whilst the other
part is precipitated with ammonia, ignited, and
weighed,—the ferric oxide being calculated by subtracting the phosphoric acid from the weight of the
precipitate, or it may be determined volumetrically
with potassic dichromate or potassic permanganate
in sulphuric-acid-solution.

Should there be an excess of phosphoric acid over
and above that requisite to combine with the iron,
then a standard solution of ferric chloride is added

until the quantity of iron present is sufficient to
combine with the whole of the phosphoric acid.
The quantity of iron added is afterwards subtracted
from that found.

b. In the filtrate the **manganese** is precipitated by
heating to 60° C. and then adding a few drops of
bromine, as described on p. 39.

The filtrate from the manganese serves, after the
addition of a little sodic acetate, for the estimation
of **lime** and **magnesia** in the manner already
described.

EXAMINATION OF PART B. — The acid-solution is
heated, and baric chloride added in as little excess as
possible ; the precipitated baric sulphate is weighed,
and from it the **sulphuric acid** calculated.  To the
filtrate ferric chloride is added, and the liquid evaporated nearly to dryness in a dish to drive off the greater
part of the free acid present ; the residue is rendered
alkaline with a solution of baric hydrate, and the whole
evaporated to dryness on the water-bath.  Extract with
water and filter ; in the filtrate precipitate the excess of
baryta together with the lime by means of ammonic
carbonate and oxalate.  Filter, evaporate the filtrate to
dryness, and ignite to expel the ammonia-salts.  Extract
the residue with dilute hydrochloric acid, filtering if necessary, and evaporate the solution of the alkaline chlorides
so obtained to dryness, and weigh.  Proceed further with
the estimation of the **alkalies** as usual, p. 19.

4. The ashes obtained after baric hydrate has been added
are treated in the manner described on p. 136, for the separation of silica, sand, and carbon.  The residue is extracted

with a solution of sodic carbonate and caustic soda. The
silica is separated from the alkaline solution, and in the
filtrate from the silica 'the sulphuric acid is precipitated
with baric chloride. The portion insoluble in alkalies is,
after treatment with dilute hydrochloric acid and washing
with water, dried and weighed ; the carbon is then burnt
off and the remainder weighed again as sand.

The acid filtrate from the sand and carbon is divided
into two parts :—

1. One part, after heating and precipitating with ammonia, ammonic carbonate, and ammonic oxalate,
   is used for the determination of the alkalies.

2. The other part, after precipitating the baryta with
   very dilute sulphuric acid, is further treated as
   described on p. 138, 3.


The following plan should be adopted if it is desired to
determine all the mineral ingredients, and to avoid
the loss of some part of them, which unavoidably occurs
when the ignition is carried on without any additions :—

1. A quantity of the substance sufficient to yield 2
   grms. of ash is carefully carbonised, and the finely-
   powdered carbonaceous mass is repeatedly extracted
   with distilled water.
   
   a. To this aqueous extract argentic nitrate is at
      once added.
      
   b. The carbonaceous residue is now extracted
      with water acidulated with nitric acid and
      thoroughly washed on a filter; the extract
      obtained is added to a.

2. The carbonaceous residue is then completely burnt
   by ignition and again extracted—first with water,

and then with hot and moderately strong nitric acid. All the above extracts are then united.

3. Determination of sulphur and chlorine.—The nitrate of silver precipitates in these extracts the sulphur of the soluble sulphur-compounds and the chlorine. The precipitated argentic sulphide and chloride are collected on a tared filter, and the filtrate united with that obtained in 4. The argentic chloride is dissolved off the filter with ammonia, and reprecipitated in the filtrate with nitric acid and weighed. The argentic sulphide is weighed, and from it the sulphuric acid calculated.

4. The portion of the ash insoluble in nitric acid is heated with concentrated hydrochloric acid and filtered. The filtrate is added to that obtained in 3, and the silver precipitated by the hydrochloric acid.

5. The portion insoluble in acids is treated as on p. 11, 1.

6. Silica, etc.—The acid filtrates free from silver are evaporated in a porcelain dish and the silica separated. The filtrate from the silica is divided into two parts—

    *A*. For ferric oxide, alumina, manganese, lime, magnesia, and phosphoric acid, see p. 105.

    *B*. For sulphuric acid and alkalies, see p. 105.

## II. EXAMINATION OF ASHES RICH IN SILICIC ACID.

If the ash is not readily decomposed by hydrochloric acid it should be treated with a solution of pure caustic soda in a platinum dish, evaporated and heated—without, however, allowing the mass to melt.

The mass is then treated with hydrochloric acid to separate the silica in the acid filtrate, from which all the other ingredients, excepting alkalies, are determined. The alkalies are determined in a special portion of the ash after fusion with baric hydrate.

## 2. ASH OF FUEL.

### I. WOOD-ASH.

The agricultural value of the ashes of wood depends chiefly upon the large proportion of **potash-salts and alkaline earthy phosphates**, whilst of secondary import- ance are the **calcic carbonate and silica** which they con- tain. As the potash-salts are almost the only ingredients of the wood-ash soluble in water, its value as regards these is generally represented by the magnitude of the aqueous extract; whilst the alkaline earthy phosphates and the car- bonates of lime and magnesia are found in the residue insoluble in water. The ash often further contains more or less sand, carbon, and clay, which are not dissolved by acids.

A considerable quantity of the ash is powdered, mixed, and then portions taken for analysis.

The complete analysis is conducted upon the same plan as that already described for the ashes of plants rich in carbonic acid, p. 136.

1. **Moisture** is determined by drying 3 grms. on a watch-glass at 100° C., until the weight is constant.

2. Treat 5 grms. of the ash with hydrochloric acid

together with a little nitric acid, evaporate to dryness, and determine in the residue insoluble in acids the **carbon, silica, sand, clay, etc.**

3. The hydrochloric-acid-solution is diluted to 500 c.c.

    *a.* Take 300 c.c. of the latter, add a little ferric chloride, and then precipitate the **ferric oxide, alumina, and phosphoric acid** by boiling with sodic acetate (see p. 42, 4, *a*). Dissolve the precipitate in nitric acid and determine the phosphoric acid with ammonic molybdate. In the filtrate from the sodic-acetate-precipitate, the **manganese, lime, and magnesia** are determined. If the estimation of iron and alumina is also of importance, then only 200 c.c. are treated as above, whilst 100 c.c. are boiled with sodic acetate without the previous addition of ferric chloride, and the precipitate obtained weighed. On subtracting from this the calculated weight of phosphoric acid, the difference represents the **ferric and aluminic oxides.**

    *b.* The remaining 200 c.c. are treated in the manner already described on p. 105, 5, for the determination of **sulphuric acid and alkalies.**

4. **Carbonic acid** is determined in 2 grms. of the ash by decomposing with dilute nitric acid according to one of the methods described on p. 60. In the solution obtained the **chlorine** is estimated by means of argentic nitrate.

5. **Volumetric estimation of alkali.**—For practical purposes exactly 6·911 grms. of ash are placed in a flask of about 300 c.c. capacity, 5 or 6 grms. of slaked lime (calcic hydrate) are added, then 40 to 50 c.c. of water, and the liquid heated to boiling. Filter into a measuring-flask of 100 c.c. capacity, and wash the residue so that the filtrate

amounts to exactly 100 c.c. Every c.c. of this solution
contains the potash from ·06911 grm. of ash. 10 c.c. of
this solution are titrated with normal nitric acid. The
number of c.c. required of the latter, multiplied by 10, gives
at once the percentage of potassic carbonate. From the
result obtained 3 c.c. have to be deducted for the nitric
acid required to neutralise the lime present in the 10 c.c.
of solution.

## II. THE ASHES OF TURF, LIGNITE, AND COAL.

1. **The ashes of turf and lignite** are sometimes rich
in **iron, decomposable silicates, and gypsum**; sometimes, on the other hand, in **calcic carbonate**; in the
former case they are red, in the latter of a lighter colour,
and often contain sand and clay. They are all poor in
potash-salts ($\frac{1}{4}$ to $\frac{1}{2}$ °/₀) and **phosphates** ($\frac{1}{3}$ to 2 °/₀), and
exhibit a much greater diversity in their composition than
the ashes of wood. Their value and application depend
mainly upon the **gypsum and carbonate** and **phosphate
of lime**—often also the **amorphous silica**—which they
contain. The examination will therefore be arranged with
regard to these constituents.

2. The **ashes of coal** are mostly remarkable for their
high percentage (50 to 80 °/₀) of **sand and clay**; they
generally also contain sulphates (3 to 10 °/₀) and sulphides of
iron and calcium, often a considerable proportion of amorphous silica, whilst the alkaline salts generally only amount
to $\frac{1}{8}$ or $\frac{1}{4}$ °/₀. A considerable quantity of fused ash is
generally admixed, so called slag or clinkers, which is very
hard, and decomposable only with difficulty. The ashes of
coal are of value as manure on account of the sulphates and

L

soluble silica which they contain ; before being applied as
such they should be sifted from the clinkers.

3. The analysis of the above ashes may generally be conducted as follows:—After carefully mixing the ash, 8 to 10
grms. are heated with hydrochloric acid, and after evaporation, separation of the silica, etc., and filtration,—

    *a.* The hydrochloric-acid-solution is examined in the
        way indicated in the Analysis of Soils.

    *b.* The residue, insoluble in acids, consisting of sand, clay,
        amorphous silica, and carbon, is further treated as in
        the examination of the Ashes of Plants, p. 136, 1.

The following results of analysis may be taken as typical
of the average composition of the ashes of these various
classes of fuel :—

| | | Ashes of | | |
| --- | --- | --- | --- | --- |
| | | Wood. | Lignite. | Coal. |
| Moisture | | 3·00 | ·95 | ·70 |
| | Potash | 4·13 | ·43 | ·22 |
| | Soda | ·32 | trace. | ·10 |
| | Sodic chloride | ·85 | | trace. |
| | Lime | 18·20 | 9·59 | ·16 |
| | Magnesia | 2·50 | 1·07 | ·12 |
| | Ferric and aluminic oxides | 5·12 | 21·00 | 2·70 |
| | Phosphoric acid | 1·53 | ·20 | trace. |
| | Sulphuric acid | 2·40 | 6·65 | ·20 |
| Soluble silica | | 8·23 | 13·50 | 4·70 |
| Carbonic acid | | 9·50 | ... | ... |
| Carbon | | 11·50 | 2·50 | 22·15 |
| Sand and clay | | 32·72 | 44·11 | 68·95 |
| | | 100·00 | 100·00 | 100·00 |

(The bracketed group from Potash to Sulphuric acid is labelled "Soluble in water and hydrochloric acid.")

## 3. DETERMINATION OF SUGAR, STARCH, INULINE, AND DEXTRINE.

All these substances are converted by treatment with dilute acids into one or other of the sugars of the glucose group (dextrose, levulose, and inverted sugar), all of which possess the general formula $C_6H_{12}O_6$ ; and since the sugars of this group can be readily estimated both by chemical, mechanical, and physical means, it is usual in the determination of the above to first convert them, then determine the sugar formed, and finally calculate the quantity of the above corresponding to the sugar found.

The sugars of the glucose group may be determined—

a. **By chemical means.** —From the quantity of cupric oxide reduced to cuprous oxide by a solution of glucose. Also from the quantity of carbonic anhydride evolved in the fermentation of the saccharine liquid.

b. **By mechanical means.** — By extracting the saccharine substance with alcohol.

c. **By physical means.**—Partly by ascertaining the specific gravity, and partly by determining the rotatory power of the saccharine liquid with the polarimeter.

### CHEMICAL METHOD (Fehling).

This method is based upon the property which a solution of glucose possesses of reducing a solution of cupric hydrate in alkaline tartrate. This solution, known as Fehling's solution, can be heated by itself without undergoing change; whilst in the presence of glucose at a temperature somewhat below boiling, it is reduced with precipitation of red cuprous oxide ($Cu_2O$). Cane-sugar has the same action, but only on boiling. As a definite weight of cupric oxide is reduced in this manner by a definite quantity of glucose, the reaction may be made the basis of a quantitative method for the estimation of glucose. One molecule of glucose ($C_6H_{12}O_6$) = 180, reduces 5 molecules of cupric oxide ($CuO$) = 397·5. Moreover, as cane-sugar, starch, inuline, and dextrine ($C_6H_{10}O_5$), are converted by treatment with dilute sulphuric or hydrochloric acids into glucose, these substances may be indirectly determined by the same method. It is only necessary to completely reduce by means of glucose a definite volume of a standard solution of cupric oxide, or to determine the weight of cuprous oxide precipitated, in order to ascertain the quantity of sugar. It is found that 100 parts of anhydrous glucose = 95 parts cane-sugar = 90 starch, inuline, or dextrine; or 1 part of cupric oxide = ·452 glucose = ·430 cane-sugar = ·407 starch or dextrine.

As many other organic substances exert a reducing action on Fehling's solution, it is necessary, if the presence of any such be suspected in a saccharine solution, to treat the liquid with basic acetate of lead or milk of lime in a measuring-flask, to fill up the latter to the mark with water, and then, after the solution has become clear, to examine with the solution of copper. If lead has been used in the clarification it is absolutely necessary that this should be removed before the liquid is titrated with Fehling's solution; this is done by adding a strong solution of sulphurous acid until the lead is completely precipitated, a little washed alumina is then added, and the liquid filtered after dilution is ready for use.

Vegetable juices and dark-coloured juices in general should be heated to boiling, a few drops of milk of lime added, and filtered through animal charcoal. The filtrate can then be used with the solution of copper.

Preparation of the solution of copper (Fehling's solution).—Exactly 34·64 grms. of pure crystallised cupric sulphate (this quantity is reduced by 5 grms. of anhydrous glucose), which has been finely powdered and well dried by pressure between blotting-paper, are dissolved in 200 c.c. of distilled water; in another vessel 173 grms. of pure crystallised Rochelle salt (sodic potassic tartrate) are dissolved in 480 c.c. of solution of pure caustic soda, 1·14 Sp. G. The two solutions are then mixed, and the clear deep-blue liquid obtained is diluted with water to 1 litre. The solution should be kept in the dark, but even then it gradually undergoes change, and it is therefore advisable only to mix the solution of copper with the Rochelle salt just before use. Thus the solution of copper may be diluted with water to 1 litre and kept in a vessel by itself, and the alkaline solution of Rochelle salt similarly diluted in another vessel. Then for use equal volumes of each must be taken; thus, if 10 c.c. of each be taken, the solution, containing ·3464 grm. of cupric sulphate, will represent exactly ·050 grm. of pure anhydrous glucose or grape-sugar.

I. Determination of Glucose or Grape-Sugar.

  a. By measurement of the saccharine solution requisite for the reduction of a given quantity of the solution of copper—

Measure 10 c.c. of the solution of copper into a porcelain

basin, dilute with about 50 c.c. of water, and heat to boiling. (If any precipitation occurs the solution is unfit for
use.) The solution of sugar (containing not more than
$\frac{1}{2}$ to 1 % of sugar) is now delivered from a burette into
the boiling Fehling solution in the dish, and the latter
stirred with a glass rod. The addition of saccharine solution is continued until the supernatant liquid in the dish
is seen to be colourless on inclining the dish a little. If
the reduction is complete, a portion of the liquid, filtered
hot, should not yield any further precipitate on being
heated up with some more sugar - solution ; nor after
acidulation with acetic acid should a brown precipitate or
colouration be produced with a drop of potassic ferrocyanide.
The titration should always be repeated a second time to
obtain greater accuracy.

Since the volume of sugar-solution required to reduce
the 10 c.c. of the solution of copper contains ·050 grm. of
anhydrous grape-sugar, the quantity of grape-sugar in the
whole saccharine liquid may be easily calculated.

### b. By weighing the cupric oxide reduced by a given quantity of the saccharine solution—

Dilute 20 c.c. of Fehling's solution with water, and heat
to boiling; now add a measured quantity of the sugar-solution, which must not be sufficient to remove the blue
colour. Continue to boil for some minutes, then filter
rapidly, and wash the cuprous oxide on the filter with hot
water until the solution runs through colourless. Dry the
filter and ignite in a weighed porcelain crucible, moisten
the residue with nitric acid, evaporate carefully, and ignite
in the oxidising flame of the lamp, then weigh the cupric
oxide and subtract the filter ash. The weight of cupric

oxide thus found, on being multiplied by ·452 (in the case of glucose), gives the weight of grape-sugar contained in the quantity of saccharine solution taken.   Or, since 1 grm. of anhydrous glucose reduces 2·205 grms. of cupric oxide, the quantity of grape-sugar may be calculated by dividing the weight of cupric oxide found by this number.

### c. By titration with Pavy's ammoniacal copper-solution—

If to Fehling's solution a sufficient quantity of ammonia be added, the cuprous oxide is not precipitated by glucose, but the deep-blue solution becomes decolourised, and the absence of any precipitate renders the end of the reaction more easily recognisable than it is when Fehling's solution is used in the ordinary way.   The ammoniacal solution of copper is, however, very readily oxidised, the blue colour being restored, so that precautions have to be taken to prevent the access of air during the operation.   This can be done by attaching the delivery-tube of the burette containing the solution of sugar to a tube passing through an india-rubber-cork, with which the flask containing the measured quantity of Pavy's solution is fitted.   Another tube passing through the same cork conducts the steam and ammoniacal vapours into a vessel of cold water ; it is advisable that this escape-tube dip into a little cup of mercury placed at the bottom of the water, to prevent the possibility of the latter running back.

Preparation of Pavy's solution.—To 120 c.c. of ordinary Fehling's solution, prepared as above, 300 c.c. of strong ammonia (Sp. G. ·880) are added, together with 400 c.c. of the caustic-soda-solution (Sp. G. 1·14).   This mixture is then made up to 1 litre.   100 c.c. of this solution are equivalent in oxidising power to 10 c.c. of Fehling's solution, or to ·050 grm. anhydrous glucose.

**Process.**—Place 100 c.c. of the above solution in a flask, together with some fragments of tobacco piping to facilitate boiling, fit on the india-rubber-cork with its tubes as above described, and heat the contents of the flask to boiling. The solution of sugar is then run in gradually from the burette into the boiling liquid, until the colour of the latter is completely discharged.

## II. ESTIMATION OF CANE-SUGAR (Saccharose).

The solution of cane- or beet-sugar (containing 1 part of the sugar to 20 parts of water), and saccharine solutions containing cane-sugar (saccharose), are "**inverted,**" *i.e.* converted into sugars of the glucose type,[1] by heating with a little dilute sulphuric acid (1 : 5).

The cane-sugar should be inverted as follows :—Dissolve 1·25 grms. of cane-sugar in 200 c.c. of water, and to this solution add 10 drops of hydrochloric acid (1·11 Sp. G.) The whole of the sugar is then inverted by heating the liquid for half an hour on the water-bath. Larger quantities of sugar must be treated with proportionately larger quantities of water and hydrochloric acid, and in the case of saccharine juices this must be done as approximately as possible. Thick mucilaginous liquids, and those which contain considerable quantities of albuminous and extract-

---

[1] "Inverted" sugar consists of equal parts of dextrose and levulose; but since the latter possesses a higher rotatory power than the former, the plane of polarisation is turned to the left by a solution of "inverted sugar," and thus it has the opposite effect to the solution of saccharose, from which it was derived. On this account the saccharose is said to be "*inverted.*"

ive matters, should be either clarified in the manner described in (1), or by means of **Graham's dialyser.**

The solution of inverted sugar prepared as above is
neutralised with sodic carbonate, and then diluted to its
original volume (200 c.c.)  The solution is now titrated
with Fehling's solution, and from the quantity of glucose
found, the cane-sugar (saccharose) is calculated.   100 parts
of glucose = 95 cane-sugar ; 1 part. of CuO = ·43 cane-
sugar ; 1 part cane-sugar reduces 2·32 CuO.


## III. ESTIMATION OF CANE-SUGAR (Saccharose) AND GRAPE-SUGAR (Dextrose).

A solution of cane-sugar only reduces Fehling's solution
when boiling, whilst the reduction is effected by the sugars
of the glucose group (dextrose, levulose, and inverted
sugar) at a temperature below boiling ; hence grape- can be
recognised in the presence of cane-sugar if the reduction
be conducted below the boiling temperature.   Nevertheless
small quantities of grape-sugar cannot with certainty be
estimated with Fehling's solution, since the prolonged
action of the alkaline liquid may invert a portion of the
cane-sugar.

When traces of grape-sugar are to be detected in the
presence of larger quantities of cane-sugar, milk-sugar, or
dextrine, it is best to make use of **Barfoed's solution.**[1]
This is prepared by dissolving 1 part of crystallised cupric
acetate in 15 parts of water, and then adding 5 c.c. of
acetic acid (38 %) to every 200 c.c. of this solution.   A
liquid containing grape-sugar gives, after boiling for a short
time with this solution and then standing, a **red precipi-**

[1] Barfoed, *Zeitschrift f. analyt. Chem. Fresen.* vol. xii. p. 27.

tate; whilst cane-sugar, milk-sugar, and dextrine are without action upon it. (2 c.c. of the sugar-solution should
be heated for one minute with ½ c.c. of Barfoed's reagent).

**Approximately** the grape- and cane-sugar may be
quantitatively determined by heating a portion of the
saccharine solution to 85° C. on the water-bath, and then
treating with Fehling's solution, as at this temperature the
cane-sugar reduces the latter only very slightly even on
standing, whilst the grape-sugar acts rapidly. After thus
determining the grape-sugar, another portion of the secchar-
ine solution is inverted by means of acid, and the total
sugar estimated in the usual way; the cane-sugar can then
be calculated by difference.

## IV. ESTIMATION OF STARCH, INULINE, AND DEXTRINE.

Inasmuch as all these substances can be completely converted into glucose by treatment with dilute acids, they
can be indirectly estimated by determining the glucose by
means of Fehling's solution. 100 parts of anhydrous glucose correspond to 95 parts of starch, inuline, or dextrine.

Inuline is the most easily converted of the three, mere
boiling with water being sufficient to transform it wholly
into levulose. The complete conversion of starch and
dextrine, on the other hand, requires the careful observance
of certain conditions respecting the dilution of the acid,
the temperature and duration of the action.

  *a.* The most rapid and complete conversion of starch
      and dextrine is effected by means of hydrochloric
      acid. Thus 1 to 1·5 grm. of starch should be placed
      in a flask with 100 c.c. of water, and after adding
      10 c.c. of hydrochloric acid (Sp. G. 1·125) the liquid

is heated for three hours on a water-bath in rapid
ebullition, an inverted Liebig's condenser being
attached to the flask. After filtering, the filtrate is
neutralised with soda and diluted to 250 c.c., the
glucose being then estimated with Fehling's solution.

b. When, as in the examination of many vegetable
substances, there is a danger of the dilute acid exert-
ing a saccharifying action upon other compounds
than the above, such as the pectines, it is advisable
to adopt the following plan. The substance should
be boiled with water until it forms a paste ; a small
quantity of a dilute acid (a few drops of 10 % sulphuric acid suffice for several grms. of starch) is
then added, and the digestion carried on at 70° C.
until a drop of the liquid gives no blue colour with
iodine, showing that the whole of the starch has
been converted into glucose and dextrine, both of
which are soluble. The liquid is now filtered, and
the filtrate diluted to a definite volume (so that 10
c.c. contain about ·1 grm. glucose). A measured
quantity of this solution is taken, and 1·5 c.c. of
dilute sulphuric acid (Sp. G. 1·113, or 160 grms. of
$SO_2(HO)_2$ per litre) added for every 10 c.c. taken,
and then digested in a sealed glass-tube for seven
hours in a paraffin-bath at 115° C. After cooling
and neutralising with plumbic carbonate, the sugar
(glucose) is determined with Fehling's solution.

c. To more completely obviate the difficulty of the
dilute acid acting upon pectine and cellulose, the
starch may be converted by means of the diastase
contained in malt. At a temperature of 65° C.
diastase has the property of converting starch into

dextrine and a peculiar kind of sugar known as
maltose, which in its properties holds an intermediate position between dextrine and dextrose.
Then, by the action of dilute acid, both the dextrine
and maltose can be converted into glucose (dextrose). For instance, 1·5 grms. of potato powder
(prepared by cutting up slices of fresh potato into
little cubes on a glass plate) are heated with 50 c.c.
of water until the starch swells into a paste, and
then 50 c.c. of a freshly-prepared **extract of malt**
are added. (The extract of malt is made by placing
100 grms. of powdered malt in a litre-flask, filling
the latter to the mark with distilled water, and then,
after standing for two hours and shaking from time
to time, the liquid is clarified by filtration through
a double filter.) The digestion with the malt extract is carried on at a temperature of 65° C. for
several hours, until the liquid gives no starch reaction with iodine. The liquid is now diluted to 250
c.c. and filtered; 50 c.c. of the filtrate are then
treated with 20 c.c. of dilute sulphuric acid (5 %),
and heated in a digester for five hours at 110 to
120° C. in a paraffin-bath, and then diluted to 250
c.c. The extract of malt is then heated in the same
way, 25 c.c. being diluted with 25 c.c. of water, and
20 c.c. of the same dilute sulphuric acid are added,
the liquid digested at 110 to 120°C., and then diluted
to 250 c.c. From each of the two solutions so prepared 50 c.c. are taken and titrated with Fehling's
solution. The 50 c.c. taken from the solution containing malt extract and original substance represent ·06 grm. substance and 2 c.c. of malt extract;

whilst the 50 c.c. from the solution containing malt extract only represent 5 c.c. of the original malt extract.  By deducting the glucose contained in 2 c.c. of original malt extract, the glucose derived from ·06 grm. of the substance can be calculated.

*d.* The following simpler method also yields very satisfactory results :—

In order to obviate the disturbing influences of hydrochloric acid acting upon substances other than the above, 1 grm. of starch (or a quantity of sub- stance containing approximately this amount of starch) is digested with 100 c.c. of water and 1 c.c. of pure hydrochloric acid (1·125 Sp. G.) at a tem- perature of 70° C. on the water-bath, until the liquid gives no starch reaction with iodine.  After filtra- tion the filtrate is diluted to 200 c.c., and 20 c.c. of hydrochloric acid (Sp. G. 1·125) are added; the whole is then digested in a flask with an inverted Liebig's condenser on a water-bath in brisk ebulli- tion for three hours to completely convert the dextrine.  When cool the liquid is neutralised with caustic soda, and after dilution to 300 c.c. titrated with Fehling's solution.

## 4. GREEN FODDER, HAY, STRAW, ETC.

The most important ingredients which must be deter- mined in order to ascertain the food-value of any of the above are—Starch, sugar, fat, albumenoids, and cellu- lose ; and in addition to these the proportion of water and ash are generally also of interest.

Sampling.—A few handfuls of the fodder are taken

from various parts of the stack; these are thoroughly mixed
on a cloth, and then cut up into pieces about 1 inch long
with a chopper.   After again mixing well, 4 to 6 lbs. are
selected in the case of hay or straw, and 8 or 10 lbs. in the
case of green fodder.

1. **Water and dry substance.**—Of the above sample
200 to 300 grms. are dried for forty-eight hours at a temperature of 60° to 70° C., and then ground finely in a small
steel mill.   The water is then estimated in 4 to 5 grms. of
this finely-divided substance by drying until constant at
a temperature of 100 to 105° C.

When great accuracy is required, the drying should be
conducted in a U-tube in a stream of dry hydrogen.

2. **Ash.**—50 grms. of the finely-divided substance are
reduced to ash in a muffle or in a platinum dish, at a dull
red-heat to avoid the loss of chlorides, in the manner
described on p. 134.   The weight of the residue represents
the **crude ash.**   In about 1 grm. of the latter the carbonic
acid and carbon are determined and deducted, the remainder being the **mineral matter free from carbonates.**
(See Analysis of Plant Ashes.)

If a further examination of the ash is not required, it is
only necessary to reduce 10 to 20 grms. of the dry substance
to ash in the first instance.

3. **Other substances.**—100 to 200 grms. of the substance, dried at 70° as above, are placed still hot upon the
mill and ground repeatedly, until the whole passes through
a sieve with meshes of 1 m.m. diameter.   This finely-ground
material is left in contact with the air for twenty-four hours,
and then preserved in well-stoppered bottles.   The quantity
of moisture must be determined in a small portion by drying
at 100 to 105° C., as before.

*a.* **Woody fibre.**—Between 3 to 6 grms. of the above air-dried material are placed in a flask and covered with 50 c.c. of dilute sulphuric acid (5 %, or 50 grms. of $SO_2(HO)_2$ in 1 litre of water); to this 150 c.c. of water are added, and the liquid boiled for half-an-hour, replacing the water as it evaporates. After subsidence the supernatant liquid is removed by means of a siphon (the shorter arm may advantageously be attached to a small funnel covered with fine muslin) into a beaker, and the residue is then twice boiled for half-an-hour with 200 c.c. of water. After removing as much of the clear liquid as possible, by decantation and siphoning, the residue, to which is added any sediment that may have become deposited in the beaker containing the acid extract, is then boiled in the same way, with 50 c.c. of dilute caustic potash (50 grms. of caustic potash in 1 litre of water) and 150 c.c. of water, and after siphoning off the clear liquid into another beaker, the residue is extracted by again boiling twice with 200 c.c. of water.

The residue left after the above process of extraction, together with any sediment deposited in the beaker containing the alkaline extract, is now thrown upon a weighed filter and washed with hot water, alcohol, and ether. (The filtration should be assisted with a pump or aspirator, as the liquid runs through very slowly.) The residue is then dried on the filter at 110° C. and weighed, and then, in order to ascertain the small proportion of ash which the woody fibre contains, it is ignited, and the ash found deducted from the previous weight. By means of a soda-lime combustion of another portion of the residue, the quantity of albumenoid matter with which the latter is mixed can be determined, and must also be subtracted before the true weight of the **woody fibre** is obtained.

The woody fibre thus obtained does not consist of pure cellulose, but of all the ligneous substances insoluble in acids, alkalies, alcohol, and ether. By the following method of Schulze's the pure cellulose can be estimated.

*b.* **Cellulose.**—Either the woody fibre obtained, as in *a*, or the original substance, may be employed for the purpose.

2 to 4 grms. of the substance are digested for twelve to fifteen days, with 12 parts of nitric acid (Sp. G. 1·10) and ·8 parts of potassic chlorate in a closed flask of large capacity. The temperature of the room in which the flask is must not exceed 15° C. When the reaction is at an end water is added and the liquid filtered, the residue is washed first with cold and then with hot water. The residue is then washed into a beaker, and there digested for about three-quarters of an hour with dilute ammonia (1 part strong ammonia to 50 of water), at a temperature of 60° C. The residual cellulose is then thrown upon a weighed filter, and washed with the above dilute ammonia until the filtrate runs through quite colourless. It is then washed successively with cold and hot water, alcohol, and ether, after which it is dried and weighed. The cellulose thus prepared always contains a small proportion (·5 %) of albu-

menoids, which must be ascertained by combustion
with soda-lime and subtracted.

c. **Fat, etc. (Ethereal Extract).**—6 to 8 grms. of the
finely-ground substance are boiled repeatedly with
ether in a flask with an inverted Liebig's condenser
attached, until the filtered extract leaves no residue
of fat on evaporation. The filtration may conveniently be conducted by having two tubes fitted
into the india-rubber-cork of the flask like the
tubes of a washbottle, the longer tube being
covered with a fine piece of muslin below. Then,
on blowing into the shorter tube, the clear ethereal
extract passes out through the longer tube.

The residue is washed on a filter with ether,
and all the ethereal filtrates diluted to a definite
volume. A fraction of this extract is then evaporated to dryness on the water-bath, and the dissolved matter determined. This will consist of
fat and chlorophyll, if the latter is contained in the
substance.

It is frequently possible to substitute for the
above the following more rapid method of extraction. The substance is placed in a wide glass-tube,
the narrower and lower extremity of which is
closed with a plug of cotton-wool, and passes
through a cork into a little flask of about 100 c.c.
capacity containing some ether, whilst the upper
extremity is attached to an inverted condenser.
The little flask is heated on the water-bath, and
the ethereal vapours pass through the cotton-wool
into the wide tube containing the substance, being
either condensed there or in the inverted con-

M

denser further on; this ether, on coming in contact
with the substance, dissolves what it can, and passes
down, filtering through the cotton-wool, into the
flask below.   This process is repeated until the
ethereal extract no longer leaves a mark when
evaporated on filter-paper.

d. Albumenoids.—Several grammes of the finely-
ground substance are dried and still further pulverised.   After drying at 100° C., about 1 grm. of
this is taken for the determination of organic
nitrogen by combustion with soda-lime.   The
albumenoid matter is then calculated by multiplying the nitrogen found by 6·25.

e. **Nutritive matter free from nitrogen.**—The
percentage of this is determined by subtracting all
the above estimations, excepting c and d, from 100.

f. **Matters soluble in water.**—The determination
of these is often of importance in feeding experiments.   The chief substances to be determined are
extractive matters, gelatine, sugar, ammonia, and
nitric acid.

a. **Total solids.**—These may be determined by
either of the two following methods :—

(1) 20 grms. of the substance roughly dried
at 70° C. are boiled for half an hour with
250 to 300 c.c. of water in a beaker, the
liquid is decanted through a filter, and the
operation repeated ten times over.   The
filtration must be accelerated with a pump,
and must not last more than a day, or else decomposition of the liquid may set in.   The
complete extraction of 15 to 20 grms. of sub-

stance in this way is usually accomplished with about 3 litres of water.

(2) By using the extraction-apparatus described on p. 161, c, using water instead of ether, it is possible to accomplish the extraction with much less liquid (500 to 700 c.c.)

In either case the aqueous extract is made up to a definite volume (say 3 litres in the first and 1 in the second case) ; 200 or 300 c.c. of this solution are evaporated to dryness in a platinum dish on the water-bath, and then placed in a drying-cupboard kept at 100° C. until the weight is constant.

β. **Mineral matters and carbonic acid.**— The residue in α is carefully incinerated and the ash determined. In about 1 grm. of the latter the carbonic acid is estimated in the manner described in the Analysis of Plant Ashes.

γ. **Soluble albumenoids.**—1000 or 500 c.c. of the aqueous extract are concentrated by evaporation in a platinum dish ; when the evaporation has nearly reached dryness, small quantities of recently ignited gypsum are added and well mixed. The mass is dried at 95° C., and a nitrogen determination made by combustion with soda-lime.

δ. **Cane- and grape-sugar, gelatine.**—About 500 to 1000 c.c. of the aqueous extract are evaporated to dryness as rapidly as possible on the water-bath, and finally under the receiver of the air-pump. The residue is extracted with boiling alcohol (85 $°/_o$) until the liquid is quite colourless. The extract is mixed with water, and the alcohol driven off by evaporation on the water-bath; in the residue so obtained the **cane-** and **grape-sugar** are determined in the manner described on p. 153, III. The portion insoluble in alcohol is dried at 100° C. and weighed, then incinerated, and the ash found subtracted from the previous weight gives the quantity of **gelatine, pectine, pectic acid,** etc.

ε. **Ammonia.**—A portion of the aqueous extract is evaporated down with a few drops of sulphuric acid; the concentrated liquor is then treated with basic acetate of lead and filtered; to the filtrate some calcic hydrate is added, and the evolved ammonia estimated with normal sulphuric acid in the manner described on p. 22, β.

ζ. **Nitric acid.**—A portion of the aqueous extract is concentrated to a few c.c., which are then introduced into Schlösing's apparatus (see p. 116).

## 5. GRAIN, FLOUR, BRAN, OILCAKE, ETC.

It is of the greatest importance that the sample be finely divided; this is accomplished by means of a little steel mill, a mortar, and a sieve with meshes only about ·001 m.m. in width. The air-dried substance must be preserved in stoppered vessels, and with this the following determinations are made:—

1. **Moisture.**—3 to 5 grms. are dried at 100° C., the loss of weight indicating the moisture.

2. **Albumenoids.**—In ·6 to ·8 grm. of the substance dried at 100° C. the organic nitrogen is determined by combustion with soda-lime. Then on multiplying by 6·25 the proportion of albumenoids is ascertained.

3. **Woody fibre, fat, and other non-nitrogenous matters** are determined in the manner already described in the Analysis of Green Fodder, Hay, Straw, etc.

4. **Ash.**—The incineration of grain is attended with more difficulty than that of hay and straw. 4 to 6 grms. of coarsely-ground substance are first carbonised in a platinum dish, and then completely reduced to ash at the lowest possible temperature.

5. **Glucose, starch, dextrine—**

  *a.* The **total quantity** of these substances is first determined by treating 1 to 1·5 grms. of the substance according to the process described on p. 157, *d.* Or 1 grm. of the dried substance may be heated with 40 c.c. of very dilute sulphuric acid (3·5 c.c. of sulphuric acid, Sp. G. 1·16 to 1000 c.c. of water) in a closed tube for eight hours, at a temperature of 140 to 150° C. in a paraffin-bath. On cooling the liquid is made up to 250 c.c. The cellulose is collected on a weighed filter, and the glucose determined in the filtrate, after neutralisation, with Fehling's solution. The cellulose is washed with water, alcohol, and ether, and then weighed.

  *b.* In a portion of an **aqueous** extract (containing, therefore, only the glucose and dextrine) prepared from 10 or 20 grms. of finely-divided substance, the glucose is determined directly by means of Fehling's solution ; whilst in another portion of the same aqueous extract the dextrine is first converted by means of dilute hydrochloric acid as in *a*, and the total glucose so obtained is then estimated with Fehling's solution.    On subtracting the glucose determined above, the difference represents the glucose-equivalent of the dextrine.

c. On subtracting the glucose = dextrine + grape-sugar found in *b* from the glucose = dextrine + grape-sugar + starch found in *a*, the difference represents the glucose-equivalent of the starch.

If the starch only is to be determined, 2 to 3 grms. of the finely-divided substance are washed on a filter—first with water, then with alcohol acidulated with sulphuric acid, and finally again with water.    By this means the dextrine and glucose are removed.    The apex of the filter-paper is then pushed through, and the residue washed into a flask, any particles of filter-paper being carefully removed.    The starch suspended in water is then converted into glucose by either of the methods described in *a*, and the latter determined in the usual way with Fehling's solution.

## 6. BEET-ROOT, TURNIPS, MANGOLDS, SUGAR-CANE, ETC.

Owing to the great development of the beet-sugar industry in France and Germany, the analysis of the beet-root has acquired a great importance on the continent. The methods employed in its examination are equally applicable to many other similar products such as those mentioned above.    When, as in this country, the chief object of the analysis is the determination of the food-value, it is necessary to ascertain, besides the proportion of sugar, also the quantity of albumenoids, cellulose, and other non-nitrogenous organic matter, together with ash, etc.

The portion insoluble in water, amounting to about 94 %, consists chiefly of cellulose and pectose. The soluble part or juice is made up principally of sugar, salts, and albumenoids.

The sugar contained in the stems of plants is almost exclusively cane-sugar (saccharose), whilst that found in their fruits is principally glucose. It is, therefore, the estimation of the cane-sugar which is of most importance in the above.

Owing to the variations in composition exhibited by different plants of the same crop, it is essential, in order to obtain results of any value, that several—about a dozen —different roots of average quality should be selected for examination. If only the composition of the tuber is required, as is the case in the beet-sugar manufacture, then the leaves above and the rootlets below should be cut off.

The tubers are first cut in half longitudinally, the one half being used for the determination of the water, the dry substance, and a few other components of the latter; whilst the other half is employed for the preparation of the juice, together with the sugar and other ingredients it contains. These latter halves are then grated to a pulp, which is placed in a linen bag, and the juice expressed by means of a hand-press.

1. Water.—From each of the above tubers cut in half a series of horizontal sections of 1 to 2 m.m. in thickness are made. These slices, weighing about 500 grms. in all, are threaded on a fine wire, weighed, and heated to 70° or 80°

C., until they are reduced to a friable mass, which is then again weighed.

This dried substance is now finely powdered, and about 5 grms. of it are taken, and the loss at 100 to 110° C. determined, from which the loss on the original substance can be calculated as water; the remainder should be preserved in well-stoppered vessels.

2. **Albumenoids.**—These are estimated by combustion with soda-lime of ·8 grm. of the **anhydrous** substance, prepared as in 1.

It is to be observed, however, that the tubers frequently contain a considerable proportion of nitric acid and ammonia, which must be determined according to the methods of Schlösing, and the ammoniacal nitrogen subtracted before the nitrogen found by combustion can be calculated into albumenoids by multiplication with the factor 6·25.

3. **The woody fibre, the ethereal extract, and ash,** are determined in the manner already described under Green Fodder, Hay, Straw, etc., p. 157.

4. **Other non-nitrogenous organic matter, sugar, etc.,** are found by difference after all the previous determinations have been made.

The **sugar** can be determined by repeated extraction of 2 to 3 grms. of the dried substance with boiling alcohol (Sp. G. ·83), in a flask fitted with an inverted condenser. The residue is dried at 100° C. and weighed; it consists of insoluble organic matters and ash, whilst the loss in weight is due almost entirely to the **sugar** dissolved. The latter is then more accurately estimated by evaporating off the alcohol and digesting the aqueous solution with a few drops of sulphuric acid on a water-bath for several hours.

The "inversion" of the cane-sugar is thereby effected, and the resulting glucose is determined with Fehling's solution. 100 parts of inverted sugar are equivalent to 95 parts of cane-sugar.

5. **Examination of the juice—**

    *a.* **Total solids.**—10 c.c. of the juice are evaporated in a little glass dish containing 20 grms. of pure ignited white sand on the water-bath, and afterwards placed over sulphuric acid in a vacuum until of constant weight.

    *b.* **Specific gravity** is ascertained by means of the specific-gravity bottle. The dry bottle of 50 c.c. capacity is weighed when empty, and then again, after being filled with the juice—great care being taken to prevent any bubbles from adhering to the sides of the glass—it is sometimes necessary to add a few drops of ether to dispel bubbles that collect in the neck of the specific-gravity bottle; the ether must be entirely absorbed with filter-paper before the stopper is placed in the bottle, and the weight taken at $17 \cdot 5°$ C. The specific-gravity bottle is afterwards weighed when full of water at the same temperature, and then, by subtracting from each of these the original weight of the empty bottle, the specific gravity of the juice can at once be calculated, since

$$\text{Sp. G. of juice} = \frac{\text{weight of juice}}{\text{weight of same volume of water.}}$$

    *c.* **Proportion of juice—**

        1. **By extraction of the pulp.**—About 20 grms. of the pulp, prepared by grating the

tuber as already described, are weighed into
a beaker diluted with from ten to fifteen times
its volume of water; after subsidence the
supernatant fluid is decanted through a
weighed filter; this process of decantation is
repeated from ten to fifteen times, and then
finally the insoluble residue is collected on
the filter and thoroughly washed there. The
residue is dried at 100 to 110° C. and weighed.

2. **By calculation.**—Let the percentage of
water in the tuber be represented by $a$, and
the percentage of water in the juice by $b$,
then

$$\text{Percentage of juice} = \frac{a \times 100}{b}$$

$d$ **Cane-sugar in juice—**

 $a$. **By means of Fehling's solution.**—Measure
25 cc. of the juice into a flask, add 5 cc. of
a saturated solution of basic acetate of lead,
and dilute up to 200 c.c. with water. Allow
to stand for ten or fifteen minutes and then
filter; measure 100 c.c. of the filtrate into a
second flask, and precipitate the excess of lead
with dilute sulphuric acid. After adding a
slight excess of sulphuric acid the liquid is
heated for one hour on the water-bath to effect
the inversion of the cane-sugar. When cool
dilute to 200 c.c. and filter; take 100 c.c. of
the clear filtrate, render it slightly alkaline
with sodic carbonate, and dilute to 200 c.c.
with water; in this manner the 25 c.c. of juice
originally taken are diluted to 600 c.c. In

this solution the inverted sugar is then determined with the alkaline solution of copper,
and the cane-sugar calculated.

β. **By means of the specific gravity of the
tuber.** — By this method, which is due to
F. Krocker, only approximately accurate
results can be obtained; but owing to the
great readiness with which the process can
be executed it is of great value where more
special appliances are inaccessible.

As a general rule the specific gravity of the
tuber increases towards its lower extremity;
and it is found that a piece of the tuber, about
equal in specific gravity to that of the whole,
is obtained if the tuber is cut by three horizontal sections into four parts of equal length,
and then the second piece from the top is
selected for examination.

In practice ten or twelve tubers are well
cleaned and cut in the manner described above.
The selected pieces are introduced into a cylindrical glass vessel of about three litres capacity
half filled with water, in every litre of which
100 grms. of common salt have been previously dissolved, and the temperature of which
should be 17° C. Fresh water is now added,
stirring the while, until an equal number of
the pieces are floating upon the surface of the
water and lying at the bottom of the vessel.
The specific gravity of the salt water is now
taken with an hydrometer, the average specific
gravity of the pieces of tuber being the same

as that of the salt water.   The following table[1]
is then applied :—

| Sp. G. of the Tuber at 17° C. | Percentage of | | Sp. G. of the Tuber at 17° C. | Percentage of | |
|---|---|---|---|---|---|
| | Sugar. | Dry Substance | | Sugar. | Dry Substance |
| 1·014 | 7·0 | 12·0 | 1·044 | 11·75 | 18·25 |
| 1·016 | 7·5 | 12·5 | 1·046 | 12·0 | 18·5 |
| 1·018 | 8·0 | 13·0 | 1·048 | 12·25 | 18·75 |
| 1·020 | 8·25 | 13·5 | 1·050 | 12·5 | 19·0 |
| 1·022 | 8·75 | 14·0 | 1·052 | 12·75 | 19·25 |
| 1·024 | 9·0 | 14·5 | 1·054 | 13·0 | 19·5 |
| 1·026 | 9·5 | 15·0 | 1·056 | 13·25 | 19·75 |
| 1·028 | 9·75 | 15·5 | 1·058 | 13·5 | 20·0 |
| 1·030 | 10·0 | 16·0 | 1·060 | 13·75 | 20·25 |
| 1·032 | 10·25 | 16·3 | 1·062 | 14·0 | 20·5 |
| 1·034 | 10·5 | 16·6 | 1·064 | 14·25 | 20·75 |
| 1·036 | 10·75 | 17·0 | 1·066 | 14·5 | 21·0 |
| 1·038 | 11·0 | 17·3 | 1·068 | 14·75 | 21·25 |
| 1·040 | 11·25 | 17·6 | 1·070 | 15·0 | 21·5 |
| 1·042 | 11·5 | 18·0 | ... | ... | ... |

γ. By the polarimeter.—The cane-sugar in
the juice, prepared as already described, can
also be determined with great accuracy and
readiness by means of the polarimeter, or
saccharimeter, as it is called when used for
this purpose.   But as the various polarimeters
(Soleil, Soleil-Ventzke, Dubosc, Wilde, etc.) are

[1] Krocker, *Leitfaden d. Agriculturchem. Analyse.*

always sold with a full description of their application, it is unnecessary to discuss the method further here.

## 7. POTATOES.

The chief constituents of the Potato are water, albumen, starch, gelatine, pectines, and woody fibre, of which the latter are especially insoluble in water. In smaller quantity the potato also contains fat, organic acids, suberous tissue in the skin, asparagine, etc.

To effect a **qualitative separation** of these substances a few potatoes should be grated to a pulp, and the latter squeezed in a linen bag with a hand-press. The juice obtained is allowed to stand until the particles of starch suspended in it have subsided. The supernatant liquid is poured off and heated to boiling; already, before the boiling temperature is reached, the coagulation of the dissolved **albumen** takes place. In the filtrate from the latter the **gelatine** and **salts** are detected.

By stirring up the residue from the press in a beaker with water, and then collecting on a hair-sieve, the greater part of the **starch** can be washed through, whilst the residue consists chiefly of **woody fibre**, but also mixed with some **pectine** insoluble in water, together with a little starch, which, owing to the imperfect comminution of the mass, remains in the cellular tissue, and can be identified with a solution of iodine.

In the **quantitative** examination of the potato, the chief determinations to be made are of water, starch, and dry substance.

1. **Water.**—About 5 or 10 grms. of finely-cut slices

of potatoes that have been previously well washed and wiped are weighed into a watch-glass or dish, and dried at 100° C. until of constant weight. The loss, usually amounting to 68 or 78 °/$_o$, is water.

More accurate results · are obtained by threading a number of thin slices of potato (either halved or whole) on a wire, and drying them at 60° C., until they can be ground in a mill. After being exposed to the air for some time, the slices are ground in a small mill, and 5 to 10 grms. of the powder are dried at 100° C. until of constant weight. The drying may with advantage be conducted in a tube through which a current of dry hydrogen is passing.

2. Starch—

    *a.* **With Fehling's solution.**—As already mentioned at another place, if the potatoes be at once treated with dilute sulphuric acid, the action of the latter upon other substances (pectine) besides starch may yield glucose, which, on determination with Fehling's solution, will make the proportion of starch appear greater than it really is. In order to obviate this the starch should be converted into glucose with extract of malt (p. 155), or by the method described at p. 166, *c.*

    In either case, median longitudinal sections should be made of the potatoes, and these cut up into minute cubes on a glass plate ; about 5 or 6 grms. of this finely-divided material are then used for examination.

    *b.* **By the specific gravity.**—The specific gravity of potatoes is found to be approximately proportional to the quantity of starch they contain, so that by determining the former it is possible to arrive at a fairly accurate conclusion about the latter. It is of great importance to select a really representative sample, the specific gravity of which can then be ascertained by one of the three following processes—

α. **With the hydrostatic balance.**—A net is attached to one of the scale-pans of a common strong balance, and exactly counterpoised. Then introduce from 16 to 20 lbs. of potatoes into the net, and weigh first in air, and then with the net and its contents completely immersed in water. Then—

$$\text{Sp. G.} = \frac{\text{weight of potatoes in air.}}{\text{loss of weight in water.}}$$

If care be taken that the water employed has a temperature of about 15° C., the results are reliable to the first two places of decimals.

β. **With salt water.**—A vessel of about 5 to 6 litres capacity is half filled with a cold saturated solution of common salt. Into this are thrown about twenty potatoes, previously well washed and dried with a cloth. Fresh water is now added, whilst stirring well, until one half of the potatoes are floating on the surface and the other half at the bottom of the vessel. The specific gravity of the brine, which is now equal to the average specific gravity of the potatoes, is taken with an hydrometer.

γ. **By the water displaced by a given**

weight.—A cylindrical vessel of about 3 litres capacity is about half filled with water, the surface of which is fixed by a mark on the glass.  About half a dozen potatoes, previously weighed, are now placed in the water, and the surface of the latter again indicated by means of a mark on the glass.  The potatoes are then removed, and the exact volume of water measured which is required to fill the vessel between the two marks.  The weight of the potatoes divided by the weight of the volume of water required gives the specific gravity of the potatoes.

After determining the specific gravity by one or other of the above methods, the following table, due to Holdefleiss. is then applied to find the corresponding percentage of starch and dry substance :—

| Specific Gravity. | Percentage of Starch. | Percentage of Dry Substance. | Specific Gravity. | Percentage of Starch. | Percentage of Dry Substance. | Specific Gravity. | Percentage of Starch. | Percentage of Dry Substance. |
|---|---|---|---|---|---|---|---|---|
| 1·070 | 14·36 | 18·02 | 1·094 | 17·59 | 22·16 | 1·118 | 22·61 | 27·75 |
| 1·072 | 14·51 | 18·27 | 1·096 | 17·97 | 22·59 | 1·120 | 23·05 | 28·23 |
| 1·074 | 14·69 | 18·54 | 1·098 | 18·36 | 23·03 | 1·122 | 23·49 | 28·71 |
| 1·076 | 14·89 | 18·83 | 1·100 | 18·76 | 23·48 | 1·124 | 23·92 | 29·19 |
| 1·078 | 15·12 | 19·14 | 1·102 | 19·16 | 23·94 | 1·126 | 24·34 | 29·66 |
| 1·080 | 15·37 | 19·46 | 1·104 | 19·58 | 24·40 | 1·128 | 24·76 | 30·13 |
| 1·082 | 15·63 | 19·81 | 1·106 | 20·01 | 24·87 | 1·130 | 25·17 | 30·59 |
| 1·084 | 15·92 | 20·17 | 1·108 | 20·43 | 25·35 | 1·132 | 25·58 | 31·05 |
| 1·086 | 16·22 | 20·54 | 1·110 | 20·86 | 25·82 | 1·134 | 25·97 | 31·50 |
| 1·088 | 16·54 | 20·92 | 1·112 | 21·30 | 26·30 | 1·136 | 26·36 | 31·94 |
| 1·090 | 16·88 | 21·32 | 1·114 | 21·74 | 26·79 | 1·138 | 26·73 | 32·37 |
| 1·092 | 17·23 | 21·74 | 1·116 | 22·13 | 27·27 | 1·140 | 27·09 | 32·79 |

The above table is only applicable in the case of healthy potatoes; the specific gravity of diseased potatoes is generally disturbed by the porous structure of the parts affected. Under such circumstances the diseased portions should be

N

carefully cut out, and the specific gravity taken of the remaining parts which are still sound.

3. **Woody fibre, albumenoids, ash, etc.**—These are determined in the same way as in beet-root, turnips, etc.

4. **Matters soluble in water.**—A weighed potato is grated over a dish until about 50 grms. of pulp have been obtained; the exact quantity taken is ascertained by weighing the potato again. The pulp, which may advantageously be mixed with powdered glass, is then washed with recently-boiled water into a percolator plugged below with cotton-wool. The mass is further extracted with recently-boiled water, and the filtration through the cotton-wool may be accelerated by forcing a stream of carbonic anhydride through the percolator. In this manner a colourless extract is obtained, which only becomes coloured on standing for some time in contact with air. The filtrate is made up to 1 litre, 200 c.c. of which are used for each of the following determinations :—

    *a.* **The solid residue and ash.**

    *b.* **Albumenoids,** by evaporating to dryness, mixing the moist residue with gypsum; after drying, the mass is burnt with soda-lime in the usual way. The albumenoids are then calculated from the nitrogen found by multiplying by the factor 6·25.

    *c.* **Grape-sugar.**—The solution is precipitated with basic plumbic acetate, and, after removal of the excess of lead by sulphuric acid, titrated with Fehling's solution.

    *d.* **Dextrine.**—The solution is concentrated by evaporation, and then 10 to 20 c.c. are treated with dilute sulphuric acid under pressure, or with hydrochloric acid, in the manner described on p.

157, *d.* The glucose is then determined, and after subtracting that found in *c*, the difference, on being multiplied by ·90, gives the quantity of dextrine.

## 8. JERUSALEM-ARTICHOKE (*Helianthus tuberosus*).

1. As far as the determination of **water, woody fibre, albumenoids,** and **fat** are concerned, the methods employed are the same as those already described in the examination of Beet-root, etc.

2. **Inuline.**—Amongst the other non-nitrogenous organic matters, the proportion of which is found by difference, after the above determinations have been made, is always a considerable quantity of **inuline** $(C_6H_{10}O_5)_n$. This inuline is always, but especially after the withering of the tubers, more or less converted into dextrose and levulose; so that together with inuline the presence of glucose must always be investigated. Sometimes the inuline appears to be more or less replaced by sugars, the nature of which is as yet but little known (synanthrose, etc.).

Since the inuline is very readily converted by dilute acids into levulose and dextrose, and inasmuch as for practical purposes the proportion of inuline is of the same value as that of glucose, it is generally only necessary to make one determination of glucose with Fehling's solution after converting with dilute acid. The glucose found is then calculated as inuline; 100 parts of anhydrous glucose being equivalent to 90 parts of inuline, or 10 c.c. of Fehling's solution to ·045 grm. of inuline.

About 5 grms. of the substance are cut into small cubes on a glass plate, and heated with 100 c.c. of water and

1 c.c. of hydrochloric acid (Sp. G. 1·125) for one hour on
the water-bath at 90° C. After filtration the solution is
treated as on p. 157, and then the glucose estimated by
titration with Fehling's solution, and calculated as inuline.

3. Pectine, gelatine, etc., are found by difference on
subtracting the inuline from the other non-nitrogenous
organic matters.

## 9. COMPLETE ANALYSIS OF TOBACCO.

To the examination of the more important vegetable
products of agricultural interest the complete analysis of
tobacco is now appended, to summarise the chief methods
of investigation which are adopted in the chemical examination of Plants.

### COMPOSITION.

The chief constituents of the tobacco plant, like those of
many other plants, are—Malic, citric, oxalic, acetic, and
pectic acids; cellulose, starch, and sugar. Substances soluble
in ether, viz.—Fat, resin, and ethereal oils; albumenoids;
further, mineral matters, which constitute about one-fifth
of the dry substance of the plant; and, finally, an alkaloid
—nicotine.

In preparing the sample about 100 leaves should be
selected and dried at 40° C.; they are then coarsely cut,
well mixed, and 100 grms. powdered as finely as possible,
and preserved in a stoppered bottle for analysis.

1. **Moisture.**—5 or 10 grms. of the finely-powdered
tobacco are dried for two hours at 100° C.; the loss in
weight represents the moisture. The chemically-combined

water is then obtained by difference after the other determinations have been made.

2. Nicotine ($C_{10}H_{14}N_2$).—This alkaloid, which is the
characteristic and active principle of the tobacco plant, is
an organic base capable of combining with two molecules
of monobasic acids.   In a state of purity nicotine is a
colourless oil, of Sp. G. 1·048, and boiling at 250° C. with

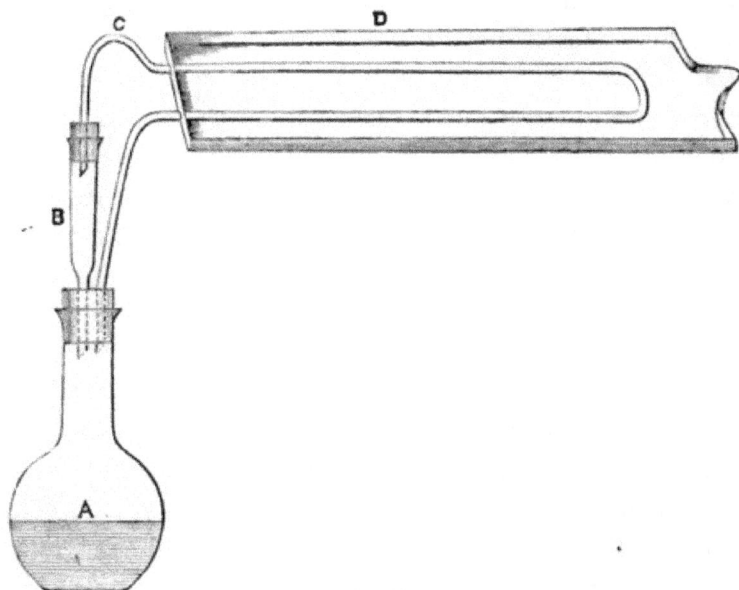

Fig. 9.

partial decomposition.   It is readily soluble in water,
alcohol, and ether.   In a current of hydrogen it distils
between 150° and 200° C. undecomposed.   In contact with
the air it rapidly acquires a brown colour.

About 10 grms. of the finely-ground tobacco are rendered
alkaline with a few drops of ammonia, in order to liberate
the nicotine from its compounds, which is then extracted
by the following apparatus :—A little flask (A, Fig. 9) of

150 c.c. capacity, about one-third filled with ether, is fitted with a doubly-perforated cork, in one of the holes of which is placed the narrower extremity of a retort adaptor, B, containing the tobacco upon a plug of cotton-wool. The wide extremity of the retort adaptor is fitted with a perforated cork bearing a long tube, C, which is twice bent in one plane, and the bend made to pass through a vessel, D, supplied with cold water; this tube, which is thus equivalent to a Liebig's condenser, then returns and passes into the second aperture of the flask. When the flask is heated the ether vapour passes through the tobacco in the adaptor, from which it extracts the nicotine, the excess of ether vapour and ammonia are condensed in the bent tube, and returned into the flask. By this arrangement the tobacco is continually extracted with an alkaline liquid, which ensures the complete removal of the nicotine. The operation is continued from four to six hours, after which the cork with its fittings is removed, and a Liebig's condenser attached. The ether and ammonia are then distilled off, until the liquid ceases to pass over with an alkaline reaction. Inasmuch as only an inappreciable quantity of nicotine passes over in this distillation at the temperature of boiling ether, the whole of the alkaloid now remains behind in the flask. The contents of the flask are then washed out into a porcelain dish with a little ether, and after spontaneous evaporation the glutinous residue, consisting of nicotine, fat, and resin, is titrated with standard sulphuric acid (5 grms. $SO_2(OH)_2$ to 1000 c.c.). One molecule of nicotine being equivalent to one molecule of sulphuric acid, thus, on multiplying the weight of sulphuric acid employed by $\dfrac{162}{96}$, the weight of nicotine is obtained.

In order to prepare nicotine in the pure state from tobacco, the leaves should be extracted with dilute sulphuric acid, and the solution, after evaporating down to half its bulk, distilled with caustic potash. The distillate is then shaken up with ether, and the ethereal extract evaporated to dryness. The residue is heated to 140° C. until free from ammonia, and then purified by distilling in a current of hydrogen at 180° C.

3. **Malic and citric acids.** — These acids, which always occur together, are present to the extent of 10 to 14 % in tobacco.

These acids cannot be separated by the ordinary method of precipitation, i.e. removing one of the ingredients by adding some reagent in excess,—as sulphuric acid, for example, is separated from nitric acid by the addition of baric chloride in excess. Recourse must therefore be had to the process of "**fractional precipitation,**" the principle of which is well illustrated by the problem in question.

The solution containing the two acids is neutralised with ammonia, and then rendered faintly acid with acetic acid. A dilute solution of plumbic acetate (4 parts of water to 1 part of a cold saturated solution of plumbic acetate) is now added, the liquid being constantly stirred all the while, until a permanent precipitate is formed (plumbic citrate is first precipitated, being less soluble in alkaline malates than plumbic malate). When the liquid has become clear, a few c.c. of the clear solution are taken up in a pipette and treated with a few drops of a very dilute solution of plumbic acetate and a drop of acetic acid ; if a further permanent precipitation takes place, it shows that the citric acid is not completely removed ;

whilst if the precipitate dissolves in the acetic acid it
shows that only malic acid remains, plumbic malate being
soluble in acetic acid.   In the first case the addition of
plumbic acetate to the original liquid is continued, small
portions being tested as above, until a precipitate is obtained which is immediately soluble in acetic acid.   The
citric acid is then known to be almost completely precipitated.

The precipitated plumbic citrate is collected on a weighed
filter, and washed with water containing a few drops of
plumbic acetate and a few drops of acetic acid (pure water
would convert the neutral plumbic citrate into insoluble
basic and soluble acid citrate, whilst water containing
plumbic acetate only would convert the neutral into basic
citrate).   The filtration must be conducted as rapidly as
possible, and finally the precipitate is washed with 36 %
alcohol until the filtrate consists of equal parts of water
and alcohol.   The presence of the alcohol causes the precipitation of the small quantity of citrate still in solution,
a small proportion of malate being also carried down.
This precipitate is also collected on a weighed filter and
washed as above; the filtrate then contains only malic
acid.

The alcohol is now evaporated off, and the residue
treated with an excess of plumbic acetate (if the alcohol
be not first removed, a basic salt of variable composition
is formed); five or six times the volume of 36 % alcohol,
containing about $\frac{1}{200}$ part of acetic acid, are now added,
and the neutral plumbic malate which now separates completely is collected on a weighed filter, after standing for
some hours, and washed as above.

The three precipitates are now dried at 100° C. until

of constant weight, and then each is detached, introduced
into a weighed porcelain crucible, and ignited; the filter
papers are ignited on the lids of the crucibles, and the ash
treated in each case with a drop of nitric acid, evaporated
to dryness, and again ignited to dull redness. The
crucibles with their lids are weighed when cool, and after
subtracting the ash the increase of weight represents oxide
of lead.

The oxide of lead from the first precipitate is calculated
into citric acid, that from the third into malic acid, whilst
that from the second may without much error be calculated
into equivalent quantities of citric and malic acids.

Citric Acid—$C_6H_8O_7$.        Malic Acid—$C_4H_6O_5$.

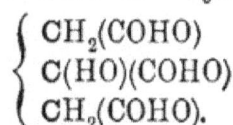

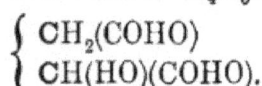

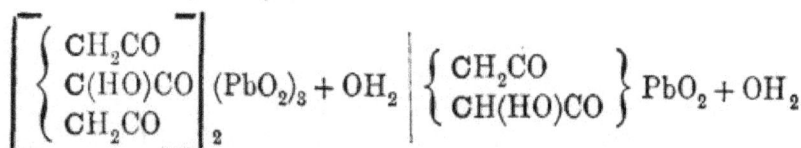

$$\begin{cases} CH_2(COHO) \\ C(HO)(COHO) \\ CH_2(COHO). \end{cases} \qquad \begin{cases} CH_2(COHO) \\ CH(HO)(COHO). \end{cases}$$

$$\left[\begin{cases} CH_2CO \\ C(HO)CO \\ CH_2CO \end{cases}(PbO_2)_3 + OH_2\right]_2 \qquad \begin{cases} CH_2CO \\ CH(HO)CO \end{cases}PbO_2 + OH_2$$

Plumbic Citrate.              Plumbic Malate.

**Extraction of malic and citric acids.**—Although
both water and alcohol dissolve these acids and their salts
from tobacco, yet, owing to the numerous other bodies
which are then obtained in the solution, it is better to
extract the tobacco with ether. In this case it is necessary
that the citric and malic acids should have been previously
liberated from their compounds by the addition of a stronger
acid, such as sulphuric acid.

In order to use as little sulphuric acid for the purpose

as possible, the ash is first examined, and from the proportions of calcic, potassic, and magnesic carbonates found the sulphuric acid required can be calculated. Twice the calculated amount of acid, diluted with five times its weight of water, is then added to 10 grms. of the tobacco in the mortar in which it was ground. The mass is well stirred in the mortar until it is of a pasty consistency. Some small pieces of pumice are then well mixed up with the mass, and the whole transferred to the retort adaptor in the extraction-apparatus described on pp. 181, 182; the mortar is finally wiped out with a piece of filter-paper which is also added. The process of extraction with ether is then commenced and carried on in the manner described in the determination of nicotine. The tartaric and oxalic acids are readily extracted in a few hours, whilst the citric and malic acids require about fifteen. The extraction may be regarded as complete when a few drops of the liquid running from the adaptor are collected and show no acid reaction.

A few c.c. of water are now added to the contents of the flask, which is then shaken to dissolve the acids which generally form oily drops round the sides of the glass. The acid liquid is then removed with a pipette, and the flask washed out with a little water, which is added to the ethereal liquid in a little beaker. The ether is now volatilised at as low a temperature as possible, leaving the aqueous solution of the acids behind.

The solution of the organic acids is usually almost colourless, and contains such a very small proportion of other substances that these may be neglected. In the case of tobacco the only acids present in the solution are acetic, oxalic, malic, citric, and tartaric acids.

**Separation of oxalic, citric, malic, and tartaric acids.**—The solution is neutralised with ammonia, the least excess of alkali being indicated by the liquid turning brown; the excess is removed by the addition of a few drops of acetic acid. The oxalic acid is now precipitated with a dilute solution of calcic acetate, an excess of the reagent being carefully avoided. The precipitate is collected on a tared filter, washed, dried at 100° C., and weighed $(C_2O_4Ca + OH_2)$; a calcium determination is further made with the precipitate in order to ascertain its purity.

The filtrate contains the **citric, malic, and tartaric acids,** which are then separated by means of plumbic acetate in the manner already described for malic and citric acids (the tartaric acid, behaving just like the citric acid, is precipitated with it). The precipitate, consisting of neutral plumbic citrate and tartrate, is filtered off, and in washing with alcohol as above a slight precipitate forms in the filtrate; this precipitate can be regarded as composed of equivalent parts of the lead-salts of the three acids. The filtrate from this, containing malic acid only, is concentrated by evaporation until the whole of the alcohol has been driven off; it is then precipitated with plumbic acetate, and after standing twenty-four hours the precipitate is filtered off and further treated as described above in the determination of malic acid.

The precipitate, containing the lead-salts of **citric and tartaric acids,** is suspended in water, and decomposed by a current of sulphuretted hydrogen. The precipitated plumbic sulphide is filtered off and washed with sulphuretted hydrogen water. The filtrate containing the free citric and tartaric acids is concentrated by evaporation; the liquid is

then divided into two equal parts, and one part very carefully neutralised with potassic hydrate, the other part is
then added (the acid potassium-salts of the two acids are
thus formed), and the whole mixed with from two to two
and a half times its volume of alcohol.   After standing for
some hours the precipitated **hydric potassic tartrate** is
collected on a tared filter, washed with alcohol, dried, and
weighed.

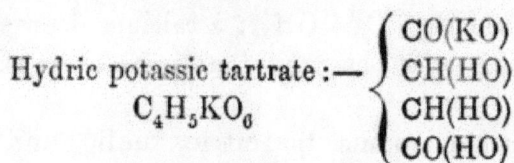

$$\text{Hydric potassic tartrate}: - \quad \begin{cases} CO(KO) \\ CH(HO) \\ CH(HO) \\ CO(HO) \end{cases}$$
$$C_4H_5KO_6$$

(Should any crystals of hydric potassic tartrate adhere
so firmly to the sides of the glass that they cannot be
detached, they should be dissolved in boiling water, and
this thrown upon the filter to dissolve the whole of the
precipitate ; the solution is collected in a tared platinum
dish, evaporated to dryness and weighed.)

The alcoholic filtrate, containing the hydric potassic
citrate, is precipitated with calcic acetate, collected on a
tared filter, and weighed as **neutral calcic citrate** :—
$(C_6H_5O_7)_2Ca_3 + 4OH_2$.

$$\begin{bmatrix} CH_2(CO) \\ C(HO)(CO) \\ CH_2(CO) \end{bmatrix}_2 (CaO_2)_3 + 4OH_2.$$

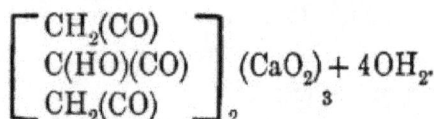

The citric acid may also be precipitated in the filtrate
from the hydric potassic tartrate with plumbic acetate, in
the manner already described in the separation of citric
and malic acids.

The acetic acid, which is also present in the extract from

the tobacco, is neglected, being determined specially as described below.

4. **Acetic acid.**—Owing to the solubility of all acetates, it is not possible to estimate acetic acid by any process of precipitation, but recourse must be had to its property of volatility with aqueous vapour; by this means it can be readily separated from the above and other fixed acids.

About 10 grms. of finely-powdered tobacco are placed in a small flask of about 200 c.c. capacity half filled with water; about ·5 grm. of argentic sulphate are now added to combine with any chlorides, and then 5 grms. of phosphoric acid. The flask is then connected with a Liebig's condenser, and the contents distilled nearly to dryness. In order to volatilise the last traces of acetic acid, water should be added and the distillation repeated. The distillate is collected in a flask and titrated with standard alkali.

The following apparatus, although more complicated, may with advantage be used for the extraction of the acetic acid :—

A large lamp-chimney (A, Fig. 10) is held in an inclined position and fitted with corks at each end. The lower and narrower cork is perforated by a glass-tube, which can be dipped into a little cup of mercury. About 10 grms. of the finely-powdered tobacco are moistened with a little water, and intimately mixed with a little finely-powdered tartaric acid. This mixture is introduced into a small retort-adaptor (B), the smaller extremity of which is plugged with cotton-wool or asbestos; the wider extremity is fitted with an india-rubber-cork and delivery-tube. This adaptor is placed inside the lamp-chimney, and its delivery-tube is made to pass through a hole in the larger and upper cork

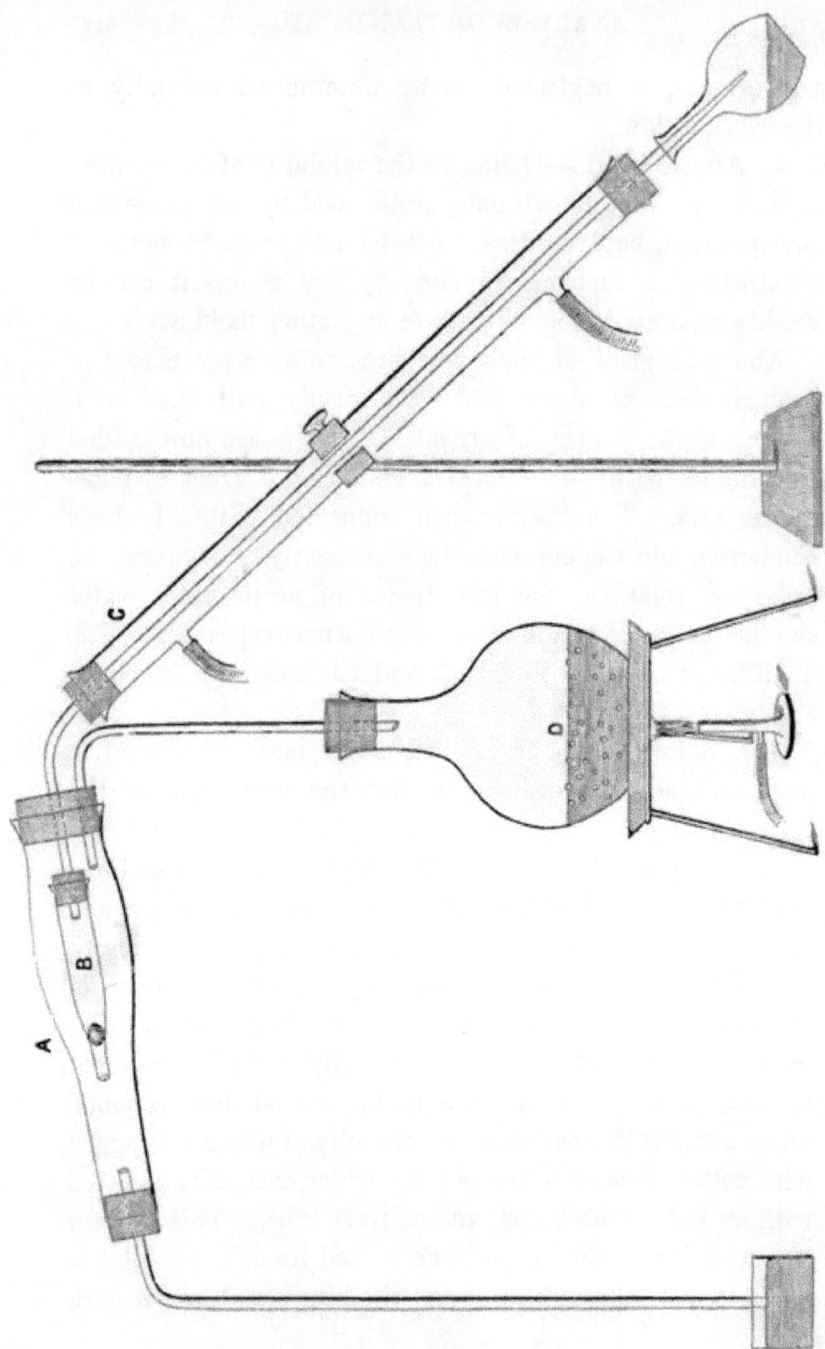

Fig. 10.

of the chimney, beyond which it is connected with a Liebig's condenser (C). The upper cork of the lamp-chimney is also perforated by a tube coming from a flask (D) containing water. The apparatus being thus put together, the water in the flask is made to boil, the steam passes into the lamp-cylinder through the cork at the upper end, and for the first few minutes it is allowed to pass out through the tube at the lower extremity. When the apparatus is **thoroughly heated** this tube at the lower extremity of the lamp-cylinder is dipped into the mercury-cup, and now the steam finds its way through the cotton-wool plug of the retort-adaptor, through the tobacco, and out by the delivery-tube and Liebig's condenser. In passing through the tobacco the steam carries with it the acetic acid, which is collected along with the condensed steam at the extremity of the Liebig's condenser. In about twenty minutes the whole of the acetic acid is thus carried over, and is determined in the distillate by titration with standard alkali. This method precludes the possibility of any hydrochloric, nitric, tartaric, or oxalic acid being carried over into the distillate.

5. **Mucilaginous matters.**—These may be grouped into three classes—**pectose**, which is neutral and insoluble; **pectine**, which is neutral and soluble; and **pectic acid**, which, although but slightly soluble itself, forms readily-soluble alkaline salts.

**Pectic acid** occurs as an insoluble lime-salt chiefly in the veins of the leaves.

The finely-powdered substance (5 grms.) is placed in a funnel and washed with 36 % alcohol, to which one-fourth its volume of concentrated hydrochloric acid has been added. When the filtrate contains no more lime, the hydrochloric acid is displaced by washing with ordinary alcohol. The residue, in which the pectic acid is now in the free state, is washed into a litre-flask, which is then three-fourths filled with distilled water, to which a hot solution of ammonic oxalate is now added; so that at least 2 grms. of oxalate are added for every 1 grm. of pectic acid present. The liquid is then kept for one to two hours at 35° C. The pectic acid dissolves in a solution of ammonic oxalate, as well as in solutions of other organic salts. The liquid is then filtered and washed, and to the filtrate an excess of calcic acetate is added. A voluminous precipitate is formed, consisting of calcic oxalate and pectate, from which the pectic acid can be separated by treatment with alcohol strongly acidulated with hydrochloric acid. If, however, the quantity of pure ammonic oxalate used be known, the precipitated lime-salts may be collected on a weighed filter, and washed first with water and then with alcohol, dried, and weighed. The precipitate is then incinerated, and the lime determined in the usual way. The lime corresponding to the ammonic oxalate used is now subtracted, and from the remainder the pectic acid can be calculated.

Pectose.—Although neither this substance nor pectine are present in tobacco, yet the methods of determining them may advantageously be described in this place.

The finely-powdered substance (5 or 10 grms.) is placed in a flask of about 250 c.c. capacity, and covered with 150 c.c. of 36 % alcohol, to which 1 grm. of ignited potassic bicarbonate, dissolved in as little water as possible, has been added. The liquid is then kept at 75° C. for half-an-hour, being frequently shaken during the time, and a cork with a glass-tube is fitted into the neck of the flask to condense the vapour of alcohol. In this operation the pectose is converted into pectic acid, whilst the presence of the alcohol prevents the conversion of the starch into sugar and dextrine. After filtration the pectic acid formed is determined in the filtrate as described above.

Pectine.—This substance is immediately converted into pectic acid by dilute alkalies, the presence of alcohol not interfering with the change. An aqueous extract of the substance is made, and by the addition of alcohol, to which a little solution of pure potassic carbonate has been added, the pectine converted into pectic acid, and the latter estimated as above.

Pectic acid and pectose.—The substance containing the two is treated with acidulated alcohol, as described under pectic acid. Pectose, not being acted upon either by acidulated alcohol or ammonic oxalate, is determined in the residue by treatment with alkaline alcohol as above.

Pectic acid, pectose, and pectine.—The pectine is determined in an aqueous extract, and the pectose and pectic acid in the residue as above. In the extraction of the pectine, however, a portion of the pectic acid passes into solution. It should then be precipitated with a neutral salt of lime, which leaves the pectine unaltered.

As a general rule it is sufficient only to estimate the total pectic acid resulting from the conversion of the other mucilaginous substances.

6. Sugar.—The leaves of fresh tobacco contain but very little sugar, whilst the pith of the stem is richer in this substance.

The tobacco is extracted with 36 % alcohol, and the solution, containing nitrates, chlorides, malates, resins, chlorophyll, nicotine, and sugar, is evaporated to dryness. The dry mass is taken up with water, filtered, and the sugar determined in the filtrate, as already described in the Analysis of Beet-root.

7. **Starch.**—This also occurs in but small quantity in tobacco. The starch is estimated, as usual, by converting into glucose with dilute acid, and then determining the latter by means of Fehling's solution. In order to obviate the error which might result from the action of the dilute acid upon the mucilaginous substances, it is necessary that these should be removed before the conversion of the starch takes place. To this end the tobacco should, as above described, be boiled with alkaline, then with acidulated alcohol, and finally with a dilute solution of ammonic oxalate. The starch in the residue is then converted into sugar with dilute hydrochloric or sulphuric acid, as described in the Analysis of Potatoes, etc.

8. **Cellulose.**—The tobacco is extracted successively with dilute sulphuric acid, caustic soda, alcohol, and ether; the residue, consisting of the woody fibre, is digested in a covered mortar with **Schweizer's Reagent** (the deep-blue solution obtained by adding ammonia to a copper-salt), the mixture being well stirred from time to time; the insoluble residue is washed with the reagent (no filter-paper must of course be used), and the solution is acidulated with acetic acid. A gelatinous precipitate of cellulose is obtained, which must be collected on a tared filter, and washed first with water and then with alcohol, after which it is dried and weighed. (See also the Analysis of Green Fodder, Hay, and Straw, p. 157.)

9. **Substances soluble in ether.**—As no satisfactory method of separating the substances soluble in ether is at present known, it is only possible to estimate the total extract, consisting of resin, wax, fat, oil, and colouring matters.

Even if the extraction with ether be long continued it is not complete, inasmuch as by a subsequent extraction with alcohol a considerable quantity of the above substances are still obtained. (See Analysis of Fodder, etc., p. 157.)

10. **Nitrogenous Compounds.** — The albumenoid nitrogen is determined by combustion of 1 grm. of the finely-powdered tobacco with soda-lime in the usual way, and then subtracting from the nitrogen found that due to ammoniacal and alkaloidal nitrogen estimated by Schlösing's method (in the cold), or by distillation with potash (see estimation of ammonia). On multiplying the difference by 6·25 (since the average quantity of nitrogen in albumenoids is 16 %) the weight of albumenoids is approximately found.

The nitrogen, as nitrates and nitrites, is estimated by Schlösing's method (see Analysis of Beet-root) with 10 grms. of tobacco; 10 grms. are also used in the above determination of ammonia.

AVERAGE COMPOSITION OF TOBACCO.

Nicotine . . . . varies from 1·5 to 9 %
    Cigars . . . „ 1·5 „ 8 „
    Havana Cigars „ 1·8 „ 2·2 „
    Smoking Tobacco „ 2·2 „ 2·5 „
    Snuff . . . „ 2 „ 3 „

Malic and citric acids (anhydrous).—10 to 14 %.
Oxalic anhydride.—1 to 2 %.
Mucilaginous substances.—5 %.
Resin, etc.—4 to 6 %.
Cellulose.—7 to 8 %.
Acetic acid.—In the fresh leaves, only in very minute

proportion; but in the species of fermentation which snuff undergoes, about 3 °/₀ of acetic acid is formed.[1]

## 10. COMPOSITION OF CEREALS AND OTHER VEGETABLE FOODS.

The following is the average composition of the cereal grains according to Graham :—

|  | Old Wheat. | Barley. | Oats. | Rye. | Maize. | Rice. |
|---|---|---|---|---|---|---|
| Water . . . | 11·1 | 12·0 | 14·2 | 14·3 | 11·5 | 10·8 |
| Starch . . . | 62·3 | 52·7 | 56·1 | 54·9 | 54·8 | 78·8 |
| Fat . . . . | 1·2 | 2·6 | 4·6 | 2·0 | 4·7 | 0·1 |
| Cellulose . . | 8·3 | 11·5 | 1·0 | 6·4 | 14·9 | 0·2 |
| Gum and Sugar | 3·8 | 4·2 | 5·7 | 11·3 | 2·9 | 1·6 |
| Albumenoids . | 10·9 | 13·2 | 16·0 | 8·8 | 8·9 | 7·2 |
| Ash . . . . | 1·6 | 2·8 | 2·2 | 1·8 | 1·6 | 0·9 |
| Loss, etc. . . | 0·8 | 1·0 | 0·2 | 0·5 | 0·7 | 0·4 |
|  | 100·0 | 100·0 | 100·0 | 100·0 | 100·0 | 100·0 |

The following numbers are taken from A. H. Church's work on *Food* :—

[1] Grandeau, *Handbuch d. Agriculturchem. Analyse.*

| | White English Wheat. | Fine Wheat Flour. | Wheat Bran. | Scotch Oatmeal. | Pearl Barley. | Rye Flour. | Cleaned Rice. | Maize. | Millet. | Dari. | Buckwheat. | Peas. | Haricot Beans. | Lentils. | Earthnuts Shelled. |
|---|---|---|---|---|---|---|---|---|---|---|---|---|---|---|---|
| Water . . . | 14·5 | 13·0 | 14·0 | 5·0 | 14·6 | 13·0 | 14·6 | 14·5 | 13·0 | 12·2 | 13·4 | 14·3 | 14·0 | 14·5 | 7·5 |
| Albumenoids and other nitrogenous bodies | 11·0 | 10·5 | 15·0 | 16·1 | 6·2 | 10·5 | 7·5 | 9·0 | 15·3 | 8·2 | 15·2 | 22·4 | 23·0 | 24·0 | 24·5 |
| Starch, with traces of Dextrine, etc. . | 69·0 | 74·3 | 44·0 | 63·0 | 76·0 | 71·0 | 76·0 | 64·5 | 61·6 | 70·6 | 63·6 | 51·3 | 52·3 | 49·0 | 11·7 |
| Fat . . . | 1·2 | 0·8 | 4·0 | 10·1 | 1·3 | 1·6 | 0·5 | 5·0 | 5·0 | 4·2 | 3·4 | 2·5 | 2·3 | 2·6 | 50·0 |
| Cellulose and Lignose . | 2·6 | 0·7 | 17·0 | 3·7 | 0·8 | 2·3 | 0·9 | 5·0 | 3·5 | 3·1 | 2·1 | 6·5 | 5·5 | 6·9 | 4·5 |
| Mineral Matter . . | 1·7 | 0·7 | 6·0 | 2·1 | 1·1 | 1·6 | 0·5 | 2·0 | 1·6 | 1·7 | 2·3 | 3·0 | 2·9 | 3·0 | 1·8 |
| | 100·0 | 100·0 | 100·0 | 100·0 | 100·0 | 100·0 | 100·0 | 100·0 | 100·0 | 100·0 | 100·0 | 100·0 | 100·0 | 100·0 | 100·0 |

|  | Potatoes. | White Turnips. | Carrots. | Beet-root, Red. | Yam. |
|---|---|---|---|---|---|
| Water . . . | 75·0 | 92·8 | 89·0 | 82·0 | 79·6 |
| Albumenoids, etc.. | 2·3 | 0·5 | 0·5 | 0·4 | 2·2 |
| Sugar . . . | ... | ... | 4·5 | 10·0 | } 16·3 |
| Starch . . . | 15·4 | ... | ... | ... | |
| Dextrine, Gum, and Pectose . . | 2·0 | 4·0 | 0·5 | 3·4 | |
| Fat . . . | 0·3 | 0·1 | 0·2 | 0·1 | 0·5 |
| Cellulose and Lignose . . . | 1·0 | 1·8 | 4·3 | 3·0 | 0·9 |
| Mineral Matter . | 1·0 | 0·8 | 1·0 | 0·9 | 1·5 |

The **soy bean** (*Soja hispida*), which is largely cultivated in both China and Japan, is the vegetable which of all others approaches most nearly in its proximate chemical composition to animal food. This will be seen from the following table, given by Prof. Kinch in his monograph on the *Soy Bean* :—

|  | Soy Bean of Japan. | Peas. | Beans. | Lupins. | Lentils. | Lean Beef. | Fat Mutton. |
|---|---|---|---|---|---|---|---|
| Water . . . | 11·3 | 14·0 | 14·8 | 12·2 | 12·5 | 72·0 | 53·0 |
| Nitrogenous Matter | 37·8 | 23·0 | 24·0 | 28·3 | 25·0 | 19·0 | 12·0 |
| Fat . . . | 20·9 | 1·7 | 1·6 | 5·0 | 1·8 | 4·0 | 32·0 |
| Carbohydrates . | 24·0 | 53·8 | 49·5 | 36·4 | 54·6 | ... | ... |
| Fibre . . . | 2·2 | 5·0 | 7·0 | 14·1 | 3·6 | ... | ... |
| Ash . . . | 3·8 | 2·5 | 3·1 | 4·0 | 2·5 | 5·0 | 3·0 |
|  | 100·0 | 100·0 | 100·0 | 100·0 | 100·0 | 100·0 | 100·0 |

# VII. ANALYSIS OF MANURES.

Owing to the introduction of artificial manures and their widespread use, the analysis of manures has of late years become of the very greatest importance; for the agricultural value of artificial manures cannot be gauged by the farmer in the same rough and ready way as that of farm-yard manures which have generally been produced on his own premises. Without a reliable analysis of the manure, the farmer is, therefore, unable to form an opinion of what is a reasonable price to pay.

The following are the ingredients which confer a value upon substances used as manure :—

1. Nitrogen in three different forms—

   a. Organic nitrogen.
   b. Nitrogen as ammonia.
   c. Nitrogen as nitric and nitrous acids.

2. Phosphoric acid in three different forms—

   a. Phosphoric acid soluble in water.
   b. Phosphoric acid soluble in ammonic citrate.
   c. Phosphoric acid soluble in acids.

3. Potash salts soluble in water—

   a. Potassic chloride.
   b. Potassic sulphate.

   *c.* Potassic carbonate.

   *d.* Potassic nitrate.

The manures of commerce may be classified as follows, according to the ingredients of agricultural value which they contain :—[1]

## CLASSIFICATION OF MANURES.

### I. NITROGENOUS MANURES.

1. Dried flesh.
2. Dried blood.
3. Cloth and woollen waste.
4. Hair, horn, and leather waste.

} Insoluble nitrogen.

5. Sulphate of ammonia.
6. Nitrate of potash (nitre).
7. Nitrate of soda.

} Soluble nitrogen and potash in nitre.

### II. PHOSPHORIC ACID MANURES.

8. Phosphorite and coprolite.
9. Precipitated phosphate of lime.
10. Bone-ash, animal charcoal.

} Phosphoric acid insoluble in water.

11. Superphosphate from phosphorite.
12. Superphosphate from animal charcoal.

} Phosphoric acid soluble in water and ammonic citrate.

### III. NITROGENOUS PHOSPHORIC ACID MANURES.

13. Bone-dust.
14. Poudrette.
15. Spent animal charcoal (from sugar refineries, etc.)
16. Fish manure.

} Phosphoric acid, organic nitrogen.

[1] Grandeau.

17. Nitrogenous superphosphates. ⎫ Phosphoric acid in
18. Guano.                        ⎬ its three forms,
19. Superphosphate guano.         ⎭ organic nitrogen,
                                    nitrogen as am-
                                    monia.

## IV. Phosphoric Acid and Potash Manures.

20. Wood-ash. ⎫ Phosphoric acid.
21. Turf-ash. ⎬ Potash.
22. Coal-ash. ⎭

## V. Potash Manures.

23. Chloride of potassium. ⎫
24. Nitrate of potash.     ⎬ Potash.
25. Carbonate of potash.   ⎭

In the following are given the determinations of importance which must be made in estimating the agricultural
value of manures :—[1]

I. Mineral    superphos - ⎫ Phosphoric acid soluble
   phates (coprolite and  ⎬ in water.
   phosphorite  treated   ⎪ Phosphoric acid soluble
   with sulphuric acid).  ⎬ in ammonic citrate.
                          ⎭ Phosphoric acid soluble
                            in acids.

II. Superphosphates  from ⎫
    bone-ash and animal   ⎬     Ditto.
    charcoal.             ⎭

[1] Grandeau.

III. Guanos treated with sulphuric acid.
- Phosphoric acid soluble in water.
- Phosphoric acid soluble in ammonic citrate.
- Phosphoric acid soluble in acids.
- Organic nitrogen.
- Ammoniacal nitrogen.

IV. Precipitated phosphate of lime.
- Phosphoric acid soluble in ammonic citrate.
- Phosphoric acid soluble in acids.

V. Bone - ash, coprolite, phosphorite.
- Phosphoric acid soluble in acids.

VI. Bone-meal, bone-turnings, animal charcoal, digested bones.
- Phosphoric acid soluble in acids.
- Organic nitrogen.

VII. Nitrate of potash.
- Nitrogen as nitric acid.
- Potash.

VIII. Nitrate of soda . . .  Nitrogen as nitric acid.

IX. Nitrate of ammonia.
- Ammoniacal nitrogen.
- Nitrogen as nitric acid.

X. Wool, waste from cloth, leather, horn, dried blood, animal refuse.
- Total nitrogen.

XI. Ashes of coal, turf, and wood.
- Total phosphoric acid.
- Potash.

XII. Potash salts, chloride of potas- $\left.\begin{array}{l}\\ \\ \\ \\ \end{array}\right\}$ Potash.
　　　sium, sulphate and carbonate
　　　of potash, distiller's refuse,
　　　etc.

**Sampling of manures for analysis.**—It is needless to say that without necessary attention being bestowed upon this point, the analysis of manures is quite valueless. The wide discrepancies which so frequently occur between the analytical results furnished by the respective chemists to the vendors and purchasers of manures are often due to the improper way in which the samples have been prepared.

In most cases the sampling is performed before the manure reaches the hands of the chemist, in which case he is not responsible for its fairness. But in all cases in which the task of preparing a representative sample devolves upon the chemist, it is necessary that he should bestow the utmost care and exercise the greatest judgment in the operation, as all the labour he may subsequently spend upon the analysis is thrown away if the sample does not really represent the average quality of the material.

The methods of analysis of a number of the more typical manures will now be described.

### 1. STABLE MANURE.

Owing to the non-homogeneous character of stable manure it is a matter of considerable difficulty to secure a representative sample.

The dung-heap should be cut through the middle, and then a vertical section of about 1 foot in thickness made from top to bottom and removed. This portion is then thoroughly mixed by itself and arranged in a square, from which a diagonal cut about 6 inches in width is taken for analysis.

In feeding experiments in which the dung is allowed to remain in the stable, the bedding is interrupted about three days before the dung is to be removed. The dung is distributed as equally as possible over the floor of the stable, and then, after removing the beasts, a diagonal cut, about 6 to 12 inches in width, is made with an axe through the dung down to the ground. This strip is then removed and weighed.

After weighing, the sample is covered with three times its volume of water and allowed to stand for a night. The fluid part is then poured through a wire-sieve, and after again pouring water on to the residual mass the particles of straw are picked out of the liquid, washed, and pressed ; the wash-water is passed through a rather fine wire-sieve, and the residue added to the particles of straw. The sample is thus resolved into

a. Wash-water.

b. Washed straw.

After stirring up the wash-water to render it homogeneous, 5 to 10 kilogrms. are taken, weighed, and filtered through fine linen or filter-paper ; the residue is dried and weighed, whilst the weight and volume are taken of the filtrate, which is then ready for examination.

The portion (b) is quickly air-dried and weighed. The straw is separated from the fæcal residue by means of a sieve, and the straw chopped up on a block. A sample is

then taken from the straw and from the fæcal residue
representing the ratio of the one to the other; a sample
of the powder filtered off from the wash-water is also taken
in proportion to the quantity of the latter, and mixed with
the above.

## A. Examination of the Aqueous Extract.

The clear fluid obtained from every 1000 grms. of
moist dung is made up to 5000 c.c., and subjected to
examination as follows:—

1. **Ammonia and organic nitrogen.**—200 c.c. (or a
quantity representing about 40 grms. of moist manure)
are supersaturated with hydrochloric acid, and evaporated down to a bulk of about 20 c.c.; in this the
**ammonia** is determined according to Schlösing's method
(p. 22).

The organic nitrogen is then determined by evaporating
down 200 c.c. to dryness (after addition of acid) on the
water-bath, and then subjecting the residue to combustion
with soda-lime in the usual way. From the nitrogen
found the ammoniacal nitrogen determined above must
be subtracted—the difference then represents the organic
nitrogen.

2. **Nitric acid.**—A quantity of solution equivalent to
100 grms. of the moist manure is treated with a little
milk of lime, and concentrated to a small bulk by evaporation on the water-bath. The lime is then precipitated
with a current of carbonic anhydride, and after filtering
the nitric acid is determined in the filtrate by Schlösing's
method with ferrous chloride (see p. 116).

3. **Sulphuric acid.**—200 c.c. of the solution are acidu-

lated with hydrochloric acid, and then precipitated with baric chloride.

4. **Sulphur compounds.** —Besides sulphates there are nearly always present unoxidised compounds of sulphur; the sulphur so combined is determined by evaporating 3 to 5 grms. of the dried residue with a concentrated solution of potassic nitrate in a platinum dish, and then gradually igniting the dried mass. The ignited residue is softened with water and heated with a little hydrochloric and nitric acids in a porcelain dish, and filtered; after separating the silica by evaporating to dryness, igniting until no more acid fumes are given off, and then taking up again with hydrochloric acid, the sulphuric acid is determined by precipitation with baric chloride. From the sulphuric acid so found that found in 3 must be deducted, the difference representing the sulphuric acid derived from the unoxidised compounds of sulphur.

5. **Total solids and ash.** —The greater part of the solution is evaporated to dryness and the residue weighed. A small portion of this residue is weighed and dried at 120° C., and from this the total solids can be calculated. This portion is then carefully ignited and the weight of the ash obtained. (The ash is further used for the determination of carbonic acid and chlorine. See Analysis of Ashes of Plants.) The remainder of the residue is then reduced to ash, the same precautions being observed as in the preparation of the ashes of plants. The mineral ingredients of the ash are determined according to the methods described in the Analysis of the Ashes of Plants, *supra*, p. 133.

## B. EXAMINATION OF THE INSOLUBLE RESIDUE.

The portion insoluble in water is air-dried, weighed, and then finely divided, if necessary, by means of a small mill.

1. **Moisture.**—About 10 grms. of the air-dried substance are dried at 110° C.

2. **Total ash.**—The above is then carefully ignited in a platinum dish and weighed. In this the carbonic acid, chlorine, and silicious particles are further determined.

3. **Composition of ash.**—A larger quantity (50 grms.) of the air-dried substance is reduced to ash at the lowest temperature possible, and the ingredients are then determined as in the Analysis of the Ashes of Plants.

4. **Organic nitrogen.**—A combustion with soda-lime is made of 1 grm. of the substance dried at 110° C.

5. **Sulphur.**—As described above, A, 4.

## 2. FÆCES.

In general the methods described in the Analysis of Green Fodder are equally applicable here.

1. **Water.**—The well-mixed substance is divided into portions of about 120 to 150 grms. (or quantities which will yield about 15 grms. of dry substance); these are placed in wide beakers, and there weighed and dried at 60° to 70° C. until the residue can be reduced to powder.

The beakers, with their contents, are weighed again in the air-dried state, and after powdering the residue in a mortar the remainder of the water is determined by drying at 100° C.

2. **Mineral matters.**—Of the dry substance obtained in 1, about 18 or 20 grms. are incinerated, and the ash examined as described in the Analysis of the Ashes of Plants, p. 133.

3. **Woody fibre.**—See the Analysis of Green Fodder, Hay, Straw, etc., p. 157.

4. **Ethereal extract.**—See the Analysis of Green Fodder, Hay, Straw, etc.

5. **Organic nitrogen.**—A soda-lime combustion is made with ·6 to ·8 grm. of the finely-powdered substance prepared in 1.

## 3. URINE.

1. **Specific gravity.**—This is ascertained by means of the specific-gravity-bottle or with an hydrometer, known as a **urinometer.**   Great care must in either case be taken to avoid the presence of froth.   The specific gravity of human urine at 15° C. varies between 1·010 and 1·040.

2. **Total solids.**—Owing to the decomposition of urea into carbonic anhydride and ammonia

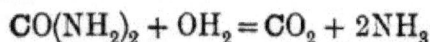

$$CO(NH_2)_2 + OH_2 = CO_2 + 2NH_3$$

when urine is evaporated, it is a matter of some difficulty to estimate the total solids it contains.   This difficulty can, however, be overcome by employing the following arrangement :—Into a small tared boat partially filled with sand 5 c.c. of the urine are measured ; the weight can then be ascertained by multiplying by the specific gravity.   This boat is then placed in a glass-tube fitted into a water-bath and open at both ends.   A stream of dry air, or, better still, of dry hydrogen, is passed through the tube heated to 100° C. by the water-bath for five or six hours.   The escaping air or hydrogen, as the case may be, is made to pass through a little flask containing a measured quantity of standard sulphuric acid. When the drying is complete the boat is weighed again in a small glass-tube in which it was previously tared. The ammonia which is evolved during the operation is absorbed by the standard acid, but there is in addition generally some ammonic carbonate, which collects on the inside of the tube dipping into the acid; this tube must therefore be carefully washed out into the standard acid, and the latter then titrated with standard alkali. The quantity of ammonia which is found to have been absorbed by the sulphuric acid is calculated into urea, and this is then added to the increase in weight of the boat.

3. **Total nitrogen.**—The residue obtained in (1) is used for combustion with soda-lime, and to the nitrogen found is added that due to the ammonia which was given off during the drying.

4. Carbonic acid—

    *a.* **Combined.**—50 to 100 c.c. are heated with a solution of pure baric chloride, and the precipitate formed is filtered off, washed, dried at 100° C., and weighed. The carbonic acid is then determined in the dried precipitate by means of Schrötter's apparatus or one of the other arrangements described in the estimation of carbonic acid.

    *b.* **Free and combined.**—50 to 100 c.c. of the urine are treated with baric chloride and ammonia, and then, after heating gently, the carbonic acid is determined in the precipitate as above.

        The free carbonic acid is found by difference from the above two determinations.

5. **Ammonia.**—The quantity of ammonia present in fresh urine is very small :—Man, ·08 to ·1 % ; cow, ·01 %.

Owing to the decomposition of the urea on distillation with alkalies, only Schlösing's method, which is conducted in the cold, is applicable to the determination of the ready-formed ammonia in urine.

6. **Urea, common salt, and chlorine.**—The urea is best determined by Liebig's method, with standard mercuric nitrate. Before this process is applicable the phosphoric acid must have been removed with baryta, and, in the case of the urine of herbivorous animals, the hippuric acid with ferric nitrate.

200 c.c. of the fresh urine are weakly acidulated with nitric acid, and, after boiling in a flask to expel the carbonic anhydride, neutralised with recently-ignited magnesia. The flask is then immersed in a vessel of cold water until the contents are cooled to the temperature of the air. The liquid is then poured into a $\frac{1}{4}$-litre-flask, so that with the rinsings it occupies a volume of about 220 c.c. The flask is then filled to the mark with a solution of ferric nitrate. The iron must be in excess, which is recognised by dipping in a piece of paper soaked in potassic ferrocyanide. Too great an excess of ferric nitrate must, however, be avoided, as it dissolves the precipitate which is at first formed. The liquid is now filtered through a **dry** filter (a wet filter is inadmissible, as the quantity of fluid must not be altered), and 150 c.c. of the filtrate are treated with a little magnesia, and made up to 200 c.c. with baryta. The liquid is again filtered through a **dry** filter, and then 15 c.c. ( = 9 c.c. of urine) of the filtrate employed for each of the following determinations :—

*a*. Determination **of common salt.**—15 c.c. of

the above are acidulated weakly with nitric acid, and then a standard solution of mercuric nitrate (for preparation of which see below) is added from a burette, until a permanent precipitate is formed. The reaction is not complete when the first opalescence appears, but the mercuric nitrate must be added until a characteristic cloudy precipitate is formed. A little practice enables the operator to reach an accuracy of ·1 c.c. The sodic chloride is then calculated as below.

Or the following process, due to Neubauer, may be adopted :—

5 or 10 c.c. of urine are evaporated on the water-bath in a little platinum dish with 1 to 2 grms. of potassic nitrate (free from chloride), then gently heated over a bunsen, and finally fused. The fused mass, free from carbon, is dissolved in a little water, rendered weakly acid with nitric acid, and then saturated with precipitated calcic carbonate. The slight excess of calcic carbonate need not be filtered off. Two or three drops of a neutral solution of potassic chromate are now added, and standard argentic nitrate then run in from a burette until a red tint is obtained. (See estimation of chlorine, p. 51.)

b. **Determination of urea.**—To 15 c.c. of the solution, prepared as above, the standard mercuric nitrate is added from a burette, a little sodic carbonate being also added each time to almost neutralise the nitric acid set free in the reaction. The point at which the urea is completely precipitated is ascertained by taking out a drop of the liquid on a glass

rod, and adding it to a drop of water rubbed up to
a paste with sodic bicarbonate on a glass plate lying
on a black sheet of paper.  As soon as the reaction
is complete the drop of liquid gives a yellow precipitate with the bicarbonate.  With practice it is
possible to add the mercuric nitrate correctly to ·1
or ·2 c.c. ( = 1 or 2 m.grm. of urea).  It is necessary that the titration should be carried on as
quickly as possible, as, even when there is no
excess of mercuric nitrate, a yellow precipitate is
obtained with the bicarbonate after standing some
time.

The quantity of mercuric nitrate required in the
determination of chlorine above, must of course
be subtracted from that used in this titration before
the urea can be calculated.

Preparation of **standard mercuric nitrate**.—An excess of **pure** mercury is digested with dilute nitric acid.
When the greater part of the mercury has dissolved, the
solution is concentrated by evaporation, and allowed to
crystallise on cooling.  The mother-liquor is poured off,
and the crystals washed with small quantities of dilute
nitric acid, and finally with cold distilled water ; then they
are dissolved in pure nitric acid, and the solution heated
until a drop produces no turbidity with a solution of common salt (showing that the whole of the mercurous salt
has been converted into mercuric nitrate).

The solution is then evaporated on the water-bath until
of a syrupy consistency ; it is then dissolved in ten times
its volume of water, and allowed to stand for twenty-four
hours, in order that any basic salt may separate, which is
then filtered off.  The filtrate is then standardised (1) by

means of a solution of **common salt** ; (2) with a saturated solution of pure **hydric disodic phosphate.**

The **solution of common salt** is prepared by dissolving 10·852 grms. of pure fused sodic chloride in 1 litre of water.

The solution of mercuric nitrate is diluted with five or ten times its volume of water, and 10 c.c. are taken and treated with 4 c.c. of the solution of hydric disodic phosphate, the whole of the mercury being precipitated as mercuric phosphate. The solution of sodic chloride is now **at once** added from a burette until the precipitated mercuric phosphate just disappears (showing that the mercuric phosphate is completely converted into mercuric chloride). Since every cc. of the sodic-chloride-solution used is equivalent to ·020 grm. of mercuric oxide (HgO), the strength of the mercurial solution can be at once calculated.

The strength of the mercurial solution required for the determination of urea is 77·2 grms. of mercuric oxide to the litre. The solution is now diluted almost to this extent, and then standardised with a solution of urea containing 2 grms. of urea dried at 100° C. in 100 c.c. By using 15 c.c. of this solution, and ascertaining the completion of the reaction by means of sodic bicarbonate, the exact strength of the mercurial solution is determined. The mercurial solution is then diluted so that 1 c.c. represents ·010 grm. of urea. The mercurial solution must then be again titrated with a solution of common salt. For this purpose 10 c.c. of a 2 % solution of sodic chloride are mixed with 3 c.c. of a 2 % solution of urea and 5 c.c. of a cold saturated solution of sodic sulphate ; the mercurial solution is then added from a burette until there is a permanent turbidity. It is then easy to calculate how much sodic chloride 1 c.c. of the mercurial solution is equivalent to. The mercurial solution is now ready for the determination of both common salt and of urea in urine.

The **solution of baryta**, employed to precipitate the phosphoric acid in urine, is prepared by mixing one volume of a cold saturated solution of baric nitrate with two volumes of a cold saturated solution of baric hydrate.

The **solution of ferric nitrate**, used for the precipitation of hippuric acid, is prepared by dissolving iron wire in nitric acid and then boiling up the solution until basic salts begin to separate. The solution, after dilution and filtration, is ready for use.

The **sodic bicarbonate** should be powdered and dry, and preserved in a stoppered bottle. Each time before use a small quantity should be washed in a watch-glass with a little cold water in order to remove the neutral carbonate, whilst crystals of bicarbonate remaining undissolved are employed as the indicator in the titration of the urea.

7. **Hippuric acid.**—200 c.c. of urine, decolourised if necessary with animal charcoal, are evaporated to one-fourth of their volume on the water-bath and then treated with 20 c.c. of hydrochloric acid, heated gently, and allowed to stand for forty-eight hours at as low a temperature as possible. The hippuric acid which crystallises out is collected on a weighed filter, and washed with a minimum of cold water until the filtrate is colourless and gives but a very slight turbidity with argentic nitrate. The filter is dried at 100° C. and weighed, and to the weight of hippuric acid found is added ·01 grm. for every 6 c.c. of filtrate. (Hippuric acid dissolves in 600 times its weight of water.) The purity of the hippuric acid may be tested by incineration, when no ash should be left.

8. **Uric acid.**—If uric acid be present (the urine of herbivorous animals contains but mere traces of uric acid, whilst that of carnivorous and omnivorous ones contains more uric than hippuric acid) it crystallises out along with the hippuric acid, from which it may be separated by digesting the mixture with 85 °/$_o$ alcohol, which dissolves hippuric acid, whilst uric acid remains almost quite un-dissolved.

9. **Sugar.**—As only grape-sugar is ever present in urine, it can be determined in the usual way with Fehling's solution (see p. 149). It is generally advisable to first decol-ourise the urine with animal charcoal; this is done by filtering two or three times through a stratum of moder-ately finely powdered animal charcoal, 24 to 30 inches in depth and 1 inch in diameter.

10. **Detection of albumen.**—A drop of acetic acid is added to 30 or 40 c.c. of urine, which is then heated to 70° C. A flocculent turbidity shows the presence of albumen.

11. **Phosphoric acid—**

    *a.* **With uranic nitrate.**—50 c.c. of filtered urine are treated with 5 c.c. of sodic acetate, and the mixture then titrated with a very dilute solution of uranic nitrate, 50 c.c. of which are equivalent to ·1 grm. phosphoric acid. (See p. 46.)

    *b.* **With magnesia-mixture.**—50 c.c. of urine are treated with magnesia-mixture, and after standing for twelve hours the liquid is filtered ; after washing the precipitate with ammonia-water (1 part strong ammonia to 3 parts water), it is dried, ignited, and weighed as magnesic pyrophosphate.

    Or the magnesia-precipitate, after washing, may be

dissolved in hot acetic acid, and the phosphoric acid determined volumetrically with uranic nitrate as above.

12. **Phosphoric acid combined with lime and magnesia.**—In order to determine the phosphoric acid combined with the alkaline earths only, 200 c.c. of urine are concentrated by evaporation, then rendered alkaline with ammonia and allowed to stand for twelve hours. The precipitate is collected, washed with ammonia-water as above, and then dissolved in a minimum of hot acetic acid; the solution is diluted to 50 c.c., and, after adding 5 c.c. of sodic acetate, titrated with uranic nitrate as above.

The lime and magnesia are determined by dissolving the phosphates, precipitated as above, in acetic acid, and then precipitating the lime with ammonic oxalate, and the magnesia, as ammonic magnesic phosphate, with ammonia, in the filtrate from the lime.

13. **Sulphuric acid.**—100 c.c. of urine are filtered, and the filtrate heated on a water-bath; after acidulation with nitric acid, baric nitrate is added, and the liquid allowed to stand for twelve hours. The precipitated baric sulphate is filtered off, washed, and ignited in a platinum crucible. After moistening with sulphuric acid it is again ignited and weighed.

14. **Sulphur organically combined.**—50 c.c. of urine are evaporated to dryness with several grms. of caustic potash and a little potassic nitrate in a silver dish. The dry mass is ignited and extracted with water, and the filtered extract precipitated with baric chloride after acidulating with hydrochloric acid. The sulphuric acid found in 13 must be subtracted, the difference being the sul-

phuric acid derived from the oxidation of the organically-
combined sulphur.

15. **Ash.**—50 c.c. of urine are evaporated to dryness
in a platinum crucible, and the residue carbonised at the
lowest possible temperature. The carbonaceous mass is
repeatedly extracted with hot water, and the residual carbon filtered off from the extract. The carbon is incinerated, and the ash added to the residue obtained on evaporating the aqueous extract; this is then ignited gently and
weighed.

The determination of the individual constituents of the
ash is carried out in the same way as in the Analysis of the
Ashes of Plants; about 6 to 8 grms. of ash, prepared as
above, are required for the purpose.


## 4. GUANO.

Guano is generally of a dark brown colour, and consists
of a light powder, with very friable lumps interspersed;
these lumps on being fractured should exhibit white spots
and crystalline structures.

As the guano is generally very far from homogeneous,
a considerable quantity should be taken and well mixed, so
as to render it of uniform composition.

According to their origin and subsequent decomposition,
or owing to additions which have been purposely made by
manufacturers, the composition of commercial guanos is
subject to great variations, more especially in the amount
of phosphoric acid and nitrogen which they contain. In
most cases the determination of phosphoric acid and nitrogen is sufficient to fix the commercial value of the guano,
but in the following the complete analysis of Peruvian

guanos is described, which may have to be slightly modified for other varieties according to the ingredients found in the qualitative examination of the guano.

1. **Water.**—2 to 3 grms. of the finely-powdered guano are dried on a watch-glass at 100° C. until of constant weight. Guano of good quality should not contain more than 14 to 18 °/$_o$ of moisture.

The escaping aqueous vapour carries with it in the above operation a little ammonia, the loss of which may even amount to more than 1 °/$_o$. If, therefore, a more accurate determination of the moisture is required, the operation of drying must be carried on in the manner described in the estimation of water in urine, in which the escaping ammonia is absorbed in standard acid. (See p. 208.)

2. **Mineral and organic matters.** — 5 grms. are weighed into a crucible, and ignited until the residue is quite incinerated, and then weighed again.

Good Peruvian guano yields about 36 °/$_o$ of a white or light-gray ash ; a much higher percentage of ash than this is indicative of adulteration, and a yellow or reddish ash point to admixture of loam, the ashes of peat, etc.

3. **Sand, etc.**—The ash obtained in 2 is heated in a beaker with hydrochloric acid, a little nitric acid, and water (much effervescence is indicative of adulteration or inferior quality), until everything but the siliceous particles are dissolved ; this residue is filtered off, washed, ignited, and weighed.

The filtrate, after separating silica by evaporation with hydrochloric acid in the usual way, is made up to 300 c.c. and employed for the determination of phosphoric acid, alkalies, and sulphuric acid.

**4. Phosphoric acid.**—It is often necessary, in the accurate determination of phosphoric acid, to fuse the guano with a mixture of dry sodic carbonate and potassic chlorate. The hydric calcic phosphate so frequently present in the different varieties of guano is, by ignition, converted into pyrophosphate, which is then only gradually reconverted into orthophosphate by solution in acids, occasioning great inaccuracy in the determination of the phosphoric acid, especially with the uranic nitrate method. There is least error when the ammonic molybdate method is employed, as the solution is then heated with nitric acid for several hours. It is therefore advantageous, in determining phosphoric acid in all guanos, to take a special portion and ignite it with two parts of dry sodic carbonate and one part of potassic chlorate. In the analysis of substances such as bone-dust, animal charcoal, etc., which on ignition yield a carbonaceous residue rich in nitrogen and combustible only with difficulty, the potassic chlorate may with advantage be replaced by potassic nitrate, which continues to give off oxygen at higher temperatures.

2 grms. of the substance are ignited with three times their weight of the above mixture in a platinum crucible at a gentle heat. As soon as the contents of the crucible are white the heat is increased, and the mass fused for a quarter of an hour at a red-heat. After cooling, the crucible is placed in a beaker and covered with 100 c.c. of water; 23 c.c. of nitric acid (1·25 Sp. G.) are then added, the beaker being partially covered with a clock-glass. The silica is separated by evaporation in the usual way, and the filtrate is made up to 300 c.c. The phosphoric acid is determined in 50 c.c. with ammonic molybdate, or by titration with uranic nitrate; in the latter case, if ferric

phosphate is precipitated from the acetic-acid-solution, it must be collected on a filter and estimated separately.

5. **Phosphoric acid, lime, magnesia, and ferric oxide.**—200 c.c. of the filtrate obtained in 3 are rendered weakly alkaline with ammonia in the cold, and the precipitate formed is dissolved in acetic acid without heating; any ferric phosphate remaining undissolved is filtered off, ignited, weighed, and the **ferric oxide** and **phosphoric acid** calculated. In the strongly-heated filtrate the **lime** is precipitated with ammonic oxalate, filtered off, and estimated. The filtrate from the lime is rendered alkaline with ammonia, and the precipitated **ammonic magnesic phosphate** filtered off and weighed, whilst the remaining phosphoric acid is precipitated in the filtrate from this with magnesia-mixture.

6. **Alkalies and sulphuric acid.**—100 c.c. of the filtrate obtained in 3 are heated nearly to boiling, and a slight excess of baric chloride added; the precipitated baric sulphate is filtered off and weighed. In the filtrate the greater part of the free hydrochloric acid is eliminated by evaporation; the liquid is then diluted and rendered alkaline with baric hydrate. The precipitate is filtered off, and the filtrate treated with ammonic carbonate and oxalate to remove the lime and excess of baryta. After filtration the alkaline chlorides are determined by evaporating the filtrate, then the potassium as potassic platinic chloride, and the sodium by difference.

7. **Carbonic acid.**—A separate portion is treated with hydrochloric acid in one of the forms of apparatus for carbonic acid determination already described.

8. **Matters soluble in water.**—5 grms. of the finely-powdered and homogeneous guano are extracted with 100

c.c. of water for a quarter of an hour, at a temperature near
the boiling point. The insoluble residue is collected on a
tared filter, dried at 100° C., and weighed.

The loss consists of—

   *a.* Moisture.

   *b.* Matters soluble in water.

If the moisture found in 1 be subtracted from this
loss, the difference gives the proportion of matters soluble
in water, which in good Peruvian guano amount to about
36 °/₀.

It must be remarked that if the above extraction with
water be continued only for a short time, the proportion
of ammonic oxalate dissolved is greater, whilst that of
ammonic phosphate and sulphate is less ; whilst if the duration of the extraction be extended, the amount of oxalic
acid dissolved diminishes and that of ammonic phosphate
increases. The small quantity of ammonic chloride and
sulphate present in the guano tends to bring the calcic
phosphate into solution, which is then decomposed by the
ammonic oxalate with precipitation of calcic oxalate and
formation of soluble ammonic phosphate.

9. **Nitrogen.**—A soda-lime-combustion is made with ·5
grm. of guano, the latter being mixed with the soda-lime in
the combustion-tube by means of a wire stirrer, and not in
an open mortar, otherwise loss of nitrogen as ammonia
would occur.

The ready formed ammonia can be estimated by
Schlösing's method.

10. **Uric acid.**—The residue insoluble in water obtained in 8 is gently heated with a dilute solution
of caustic soda, and filtered ; the filtrate is then acidulated with hydrochloric acid. The precipitated uric acid

is collected on a tared filter, dried at 100° C., and weighed.

**11. Oxalic acid.**—The carbonic acid is expelled from a separate portion of the guano by treatment with dilute sulphuric acid; the latter is then neutralised with caustic soda (free from carbonate), and then mixed with pure peroxide of manganese. The whole is then introduced into a carbonic-acid-apparatus, and sulphuric acid added; the oxalic acid is decomposed with evolution of carbonic anhydride, the quantity of which is determined by loss in the usual way. From the carbonic anhydride evolved the oxalic acid is then calculated.

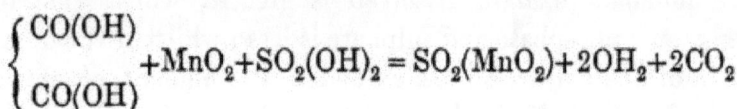

$$\left\{ \begin{array}{l} CO(OH) \\ \\ CO(OH) \end{array} \right. +MnO_2+SO_2(OH)_2 = SO_2(MnO_2)+2OH_2+2CO_2$$

## 5. BONE-DUST, BONE-ASH, ANIMAL CHARCOAL.

**A. Bone-Dust.**—It is necessary in estimating the agricultural value of bone-dust to take its mechanical condition as well as chemical composition into consideration. As regards its chemical composition, it is generally sufficient to determine the proportion of phosphoric acid and nitrogen in order to gauge its commercial value.

In practice a distinction is made between steamed and unsteamed bone-dust. The former is prepared by pulverising bones which have been extracted with steam, whilst the latter is prepared from more or less fresh bones. Owing to the difficulty of pulverising the fresh bones, the latter class of bone-dust is generally less finely divided than the former.

## CHEMICAL EXAMINATION.

1. **Water.**—3 grms. of the finely-powdered bone-dust are dried on a watch-glass at 100° C. until of constant weight.

2. **Organic matter and ash.**—3 grms. of the finely-divided bone-dust are placed in a platinum crucible, and ignited at first very gently, and finally at a red-heat, until the ash appears quite white. The ash is treated with a solution of pure ammonic carbonate, and dried at 150° C., and then weighed. The loss, after subtracting the moisture found in 1, represents the **organic matter**; the weight of the residue gives the **ash** or **mineral matter**.

3. **Sand, insoluble matters.**—The ash obtained in 2 is treated with a little nitric acid and washed into a beaker, in which it is heated until all has dissolved, excepting a residuum of **sand, etc.** The liquid is filtered, and the residue on the filter washed with water until the fluid passing through ceases to redden litmus-paper. The filtrate is made up to 250 c.c. The residue is dried, ignited, weighed, and calculated as **insoluble matter**.

4. **Phosphoric acid** (volumetrically).—50 c.c. of the filtrate obtained in 3 are titrated with uranic nitrate in the manner described on p. 46.

5. **Nitrogen.**—A soda-lime-combustion is made with ·6 to ·8 grm. of the bone-dust in the usual way.

The above determinations are sufficient to judge of the purity of the bone-dust; the following may, however, sometimes also be of interest:—

6. **Phosphoric acid** (gravimetrically) **and alkaline earths, etc.**—3 grms. of the bone-dust are incinerated, and the ash dissolved in dilute hydrochloric acid; after separat-

ing the insoluble portion as above, the filtrate is made up to
250 c.c.

  *a.* **Phosphoric acid, lime, magnesia, ferric oxide.**
  —100 c.c. of the acid-solution are treated with a
  slight excess of ammonia in the cold; then acetic
  acid is added until the phosphates of the alkaline
  earths, precipitated by the ammonia, are redissolved.
  Any ferric phosphate remaining undissolved is filtered off, dried, ignited, and weighed, the **phosphoric acid** and **ferric oxide** being calculated.

  The filtrate is heated, and ammonic oxalate added;
  the precipitated calcic oxalate is filtered off, dried,
  and ignited; and the calcic carbonate or oxide so
  obtained weighed, and the **lime** calculated.

  The filtrate from the lime is divided into two equal
  parts. In one part the **phosphoric acid** is precipitated with magnesia-mixture after rendering
  alkaline with ammonia. The precipitated ammonic
  magnesic phosphate is collected on a filter and treated
  in the usual way. In the other part the **magnesia**
  is precipitated with hydric disodic phosphate as
  ammonic magnesic phosphate, which is collected
  on a filter and treated in the usual way.

  *b.* **Alkalies and sulphuric acid.**—With the remaining 150 c.c. of the acid-solution, in the same
  way as described in the Analysis of Guano.

7. **Carbonic acid** (calcic carbonate).—Three grms. of
the bone-dust are treated with hydrochloric acid in a
carbonic-acid-apparatus (see p. 60), and the loss due to
evolution of carbonic anhydride determined; from this the
carbonic acid and its equivalent of calcic carbonate can be
calculated.

## MECHANICAL EXAMINATION.

The agricultural value of bone-dust depends not only upon its chemical composition, but also upon the state of division in which it is employed as a manure. The more finely divided it is the more readily can it be mixed with the soil, and the more rapidly does it there become decomposed.

The state of division can be measured by weighing the respective quantities of bone-dust which pass through sieves of different degrees of fineness.

Three sieves may conveniently be used, 50 or 100 grms. of the bone-dust being sifted :—

    *a.* **Very fine** passes through No. 1 sieve with about 4000 meshes to the square inch.

    *b.* **Fine** passes through sieve No. 2 with 2000 meshes to the square inch.

    *c.* **Tolerably fine** passes through sieve No. 3 with 1000 meshes to the square inch.

    *d.* **Coarse.**—The residue which does not pass through sieve No. 3.

The following is given by Krocker as the average of a large number of analyses of bone-dust—

## A. CHEMICAL COMPOSITION.

|  | Steamed. | Unsteamed. |
|---|---|---|
| Moisture . . . | 5·30 | 7·50 |
| Organic matter . . | 33·40 | 38·00 |
| Phosphoric acid . | 22·80 | 19·50 |
| Lime . . . . | 27·70 | 24·20 |
| Carbonic acid . . | 3·80 | 4·10 |
| Ferric oxide . . . | ·90 | 1·20 |
| Magnesia, Alkalies, Sulphuric acid, etc. | 2·50 | 2·00 |
| Insoluble matters . . | 3·60 | 3·50 |
|  | 100·00 | 100·00 |
| Nitrogen . . . | 3·80 % | 4·05 % |

## B. MECHANICAL COMPOSITION.

|  | Steamed. |  |  |  | Unsteamed. |  |  |
|---|---|---|---|---|---|---|---|
| I. Very fine . | 45 | 56 | 65 | 75 | 20 | 31 | 35 |
| II. Fine . . | 12 | 20 | 15 | 12 | 8 | 9 | 15 |
| III. Tolerably fine | 16 | 18 | 12 | 9 | 14 | 18 | 14 |
| IV. Coarse . . | 27 | 6 | 8 | 4 | 58 | 42 | 36 |
|  | 100 | 100 | 100 | 100 | 100 | 100 | 100 |

B. BONE-ASH.—The analysis of the mineral matter in bones is conducted in the manner already described under Bone-dust.

C. ANIMAL CHARCOAL—

    *a.* **Moisture.**—3 to 5 grms. of the finely-powdered and well mixed substance are dried at 100° C. until of constant weight.

    *b.* **Carbon, sand, etc.** — 3 grms. of the animal charcoal are digested for one hour with 50 c.c. of water and 10 c.c. of strong nitric acid on the water-bath. The residue of carbon, sand, etc., is collected on a tared filter, dried at 100° C., and weighed; it is then ignited in a platinum crucible until the carbon is burnt off, and then weighed again. The residue is calculated as sand, etc., the loss as carbon. The filtrate is made up to 250 c.c. and used for the following determinations.

    *c.* **Phosphoric acid.**—50 c.c of the filtrate obtained in *b* are treated with 150 to 200 c.c. of ammonic-molybdate-solution, and the determination of the phosphoric acid conducted as described on p. 44.

    *d.* **Sulphuric acid and chlorine.** — These are determined in separate portions of the remainder of the filtrate obtained in *b*; the sulphuric acid being precipitated with baric chloride, the chlorine with argentic nitrate.

    *e.* **Ferric oxide, lime, magnesia, etc.**—By incinerating 3 grms. of the substance and proceeding as described in the Analysis of Bone-dust.

    *f.* **Carbonic acid.**—By decomposing 1 to 2 grms. in a carbonic-acid-apparatus with dilute hydrochloric or nitric acid (see p. 60). The loss of sulphuretted hydrogen in the same operation can be obviated by previously adding a little cupric sulphate.

*g.* **Calcic hydrate,** which is generally present in small quantity, is determined by moistening a weighed amount of animal charcoal with ammonic carbonate in a platinum crucible, then evaporating to dryness and repeating the operation. Finally, the mass is more strongly heated with the crucible lid on, actual ignition and combustion being carefully avoided. The carbonic acid in the mass so treated is determined, and by subtracting the carbonic acid found in *f,* the carbonic acid corresponding to the calcic hydrate contained in the animal charcoal is obtained.

The following is given by Krocker as the average composition of animal charcoal :—

| | |
|---|---:|
| Moisture | 2·350 |
| Carbon and volatile matters | 12·388 |
| Lime | 38·416 |
| Phosphoric acid | 29·690 |
| Carbonic acid | 2·400 |
| Sand | 13·300 |
| Other matters | 1·456 |
| | 100·000 |

## 6. PHOSPHORITE, COPROLITE, ETC.

Owing to the large quantity of ferric oxide which these phosphatic manures frequently contain, it is only possible to determine the phosphoric acid with accuracy by means of the molybdate method.

For this purpose ·5 grm. of the finely-divided and well-

mixed substance is treated with 10 c.c. of strong hydrochloric acid in a porcelain dish, and evaporated to dryness
on the water-bath. The residue is moistened with about
2 c.c. of hydrochleric acid, and after some time 10 c.c. of
nitric acid (Sp. G. 1·2) are added. The liquid is filtered,
in order to separate any insoluble matter, and the filtrate
evaporated almost to dryness. After the addition of 5 c.c.
of nitric acid, the liquid is washed into a beaker with
a little water, and then treated with 150 to 200 c.c.
ammonic-molybdate-solution, and the determination continued as usual.

Another more rapid method, which yields tolerably
accurate results, is to ignite 5 grms of the powdered
phosphorite in a closed platinum crucible until the mass
cakes together. By this means the ferric oxide is rendered
insoluble. The ignited mass is now mixed in a mortar to
a stiff paste, with 10 c.c. of dilute sulphuric acid (5 %) ;
more acid (about 100 c.c.) and water are then added, and
the whole is washed into a ¼-litre-flask and allowed to
stand four hours. After making up to the ¼-litre and
filtering off any insoluble residue that may be present, 50
c.c. of the filtrate (1 grm. phosphorite) are treated with 5
drops of a saturated solution of pure citric acid ; the
solution is then rendered alkaline with caustic soda, and
the precipitate formed dissolved in acetic acid ; 10 c.c. of
a strong solution of sodic acetate are now added, and the
liquid titrated with standard uranic nitrate.

The following is the composition of a rich Spanish phosphorite recently analysed by me :—

| | | |
|---|---|---|
| Moisture | . | 1·76 |
| Organic matter | . | 4·09 |
| Silica | . | 1·14 |
| Sulphuric acid | . | ·25 |
| Carbonic acid | . | 3·33 |
| Phosphoric acid | | $36·52 = 79·72$ % $P_2O_2(CaO_2)_3$ |
| Lime | . | 45·05 |
| Magnesia | . | ·11 |
| Alumina | . | 5·27 |
| Ferric oxide | . | 1·80 |
| Soda | . | ·21 |
| Potash | . | ·77 |

$$\overline{100·30}$$

## 7. SUPERPHOSPHATES.

Neutral calcic phosphate is decomposed by sulphuric acid according to the equation :—

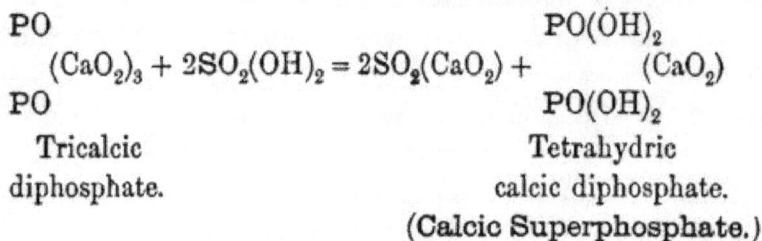

$$\begin{array}{ccc} PO & & PO(\dot{O}H)_2 \\ (CaO_2)_3 + 2SO_2(OH)_2 = 2SO_2(CaO_2) + & & (CaO_2) \\ PO & & PO(OH)_2 \end{array}$$

| Tricalcic | Tetrahydric |
|---|---|
| diphosphate. | calcic diphosphate. |
| | (Calcic Superphosphate.) |

The **superphosphate** so formed possesses a greater agricultural value than the neutral phosphate, owing to its solubility in water, and the greater rapidity with which it consequently becomes available for plant-nutrition. The treatment of insoluble phosphates with sulphuric acid, with a view to converting them into superphosphates, is the most important operation in the manufacture of artificial manures.

If the conversion into superphosphate be incomplete, a secondary reaction will take place between the superphosphate and the undecomposed neutral phosphate :—

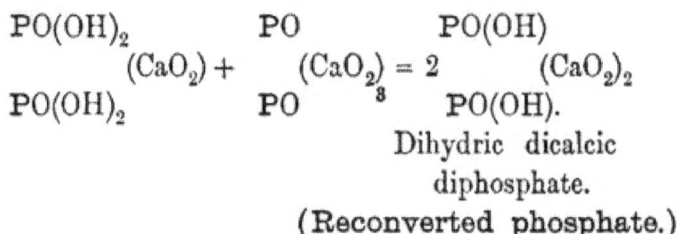

$$\begin{array}{l} PO(OH)_2 \\ \qquad (CaO_2) + \\ PO(OH)_2 \end{array} \quad \begin{array}{l} PO \\ \quad (CaO_2) \\ PO \end{array}_3 = 2 \quad \begin{array}{l} PO(OH) \\ \qquad (CaO_2)_2 \\ PO(OH). \end{array}$$

Dihydric dicalcic<br>diphosphate.<br>(Reconverted phosphate.)

The tendency to form this reconverted phosphate is greater the more iron and alumina the original phosphate contains ; thus phosphorite and coprolite which have been treated with sulphuric acid are very prone to form reconverted phosphate. Reconverted calcic phosphate is insoluble in water, but soluble in a solution of **ammonic citrate** and other organic salts. Since this reconverted phosphate is of greater value as a manure than the neutral calcic phosphate, but inferior to the superphosphate, it is often necessary, in the analysis of superphosphate manures, to determine the amount of phosphoric acid in each of these three states of combination.

Superphosphates are, according to the raw materials from which they are prepared, or to additions which may have been made during their manufacture, either nitrogenous (ammonia, nitric acid, or organic nitrogen) or non-nitrogenous, or they may contain potash. The determination of these ingredients is, therefore, necessary in their analysis.

The sample for analysis should be prepared from a large quantity of the manure, which is rendered homogeneous by sifting out and pulverising the larger pieces.

1. **Moisture and chemically-combined water.**—

3 grms. of the finely-powdered substance are dried on a watch-glass at 100° C.; the loss represents the accidental **moisture.**

The residue is then dried at 150° C. until it ceases to lose weight. This further loss in weight is calculated as **chemically-combined water.**

2. **Organic and volatile matter.**—The organic matter cannot be determined by direct ignition, as on heating superphosphates sulphuric acid is volatilised. 3 grms. of the substance are, therefore, rendered alkaline with milk of lime in a platinum crucible, and the mass well stirred with a glass rod. After carefully washing the glass rod with a little water, the moisture is evaporated, and the mass dried at 150° C. and weighed; it is then ignited and weighed again. The loss is calculated as organic matter. As, however, many organic substances are decomposed at 150° C., it is often preferable to dry the mass at 100° C., and then to ignite, calculating the organic matter and chemically-combined water together.

3. **Insoluble mineral matter** (sand, clay, etc.)—The residue obtained in (1) is heated with hydrochloric acid and water; the insoluble part is filtered off, washed, dried, ignited, and weighed.

4. **Phosphoric acid—**

*A.* In superphosphates which contain little or no ferric oxide and alumina, such as the superphosphates prepared from guano (Baker, Mejillones, and Peruvian), animal charcoal, and bone-dust.

    *a.* Phosphoric acid **soluble in water—**

        α. Volumetrically.—20 grms. of the finely-powdered sample are triturated with a little water in a mortar, the insoluble particles are allowed to subside, and the supernatant liquid poured into a litre-flask. Finally, the insoluble portion is also washed into the flask, and the latter three-fourths filled with water. The flask is set aside for two hours, during which time it is frequently shaken. The flask is now filled to the mark with water and filtered ; 50 c.c. of the filtrate (corresponding to 1 grm. of substance) are used for titration with uranic nitrate ; dilute caustic soda is added until the reaction is slightly alkaline, then acetic acid until the acid reaction is restored, the titration with standard uranic nitrate being then performed as usual. (See p. 47.)

If in the preparation of the above acetic-acid-solution any considerable precipitation of ferric phosphate takes place, then 200 c.c. of the solution should be taken, and the **ferric phosphate** precipitated in the cold as above. The precipitate is collected on a filter, washed, dried, ignited, and weighed, and the **iron** and **phosphoric acid** calculated. The filtrate is made up to a definite volume, and a portion taken for titration with uranic nitrate in the usual way. When there is a very heavy precipitate of ferric phosphate, the phosphoric acid should be determined by the molybdate method.

$\beta$. **Gravimetrically.** — 50 c.c. of the above aqueous extract are treated in the manner described in the Analysis of Bone-dust.

*b.* **Insoluble phosphoric acid.** — 2·5 grms. of the finely-powdered substance are heated on the water-bath with 50 c.c. of water and 10 c.c. of nitric acid (1·2 Sp. G.) in a 250 c.c. flask until dissolved. On cooling, the liquid is made up to 250 c.c. and, after shaking up, filtered. In 50 c.c. of the filtrate (corresponding to ·5 grm. of substance) the total phosphoric acid is determined as above, either volumetrically or gravimetrically. On subtracting the amount of soluble phosphoric acid found in *a*, the difference gives the phosphoric acid insoluble in water.

*B.* In superphosphates which, like those prepared from phosphorite, contain **a large quantity of iron,** it is necessary to prepare the solution otherwise, owing to the tendency to decomposition.

*a.* **Soluble phosphoric acid.**—10 grms. of the substance are triturated with a little cold water in a mortar. After subsidence the supernatant liquid is poured through a filter into a litre-flask, and the extraction is repeated until the liquid runs through without an acid reaction. The filtrate is made up to 1 litre, and in 50 c.c. the phosphoric acid is determined with ammonic molybdate.

*b.* **Insoluble phosphoric acid**—2 grms. of the superphosphate are treated in the manner described in the determination of phosphoric acid in phosphorite. From the phosphoric acid found that obtained in *a* is subtracted, the difference being the total phosphoric acid insoluble in water, including neutral phosphate and reconverted phosphate.

*c.* **Reconverted or "reduced" phosphoric acid.**—

2 grms. of the superphosphate are stirred in a porcelain or glass mortar with a solution of ammonic
citrate (1·09 Sp. G.)  After the coarser particles
have subsided, the supernatant liquid is poured off
into a little flask, the residue is reduced to a paste
by means of a pestle, and is then also washed into
the little flask with the same solution, of which 100
c.c. in all are used.  The contents of the flask are
now heated to a temperature of 30 to 40° C. for
half-an-hour, shaking frequently during this time,
and then filtered.  The residue is repeatedly washed
with a mixture of equal parts of water and the above
solution of ammonic citrate.  The filtrate is evaporated to dryness in a platinum dish, then ignited
more strongly with sodic carbonate, and if necessary with the addition of a little potassic nitrate.
The ignited mass is dissolved in nitric acid, and the
phosphoric acid precipitated with ammonic molybdate.  From the phosphoric acid found the soluble
phosphoric acid obtained in *a* must be subtracted,
the difference being the reconverted phosphoric acid.

5. Potash, lime, etc.—
　　*a.* Potash. — 5 grms. of the superphosphate are
　　treated with 50 to 60 c.c. of water in a 250 c.c. mea-
　　suring-flask ; heat to boiling, and when cool dilute
　　up to the mark, and filter.  50 c.c. of the filtrate
　　are heated to boiling, and then baric chloride added
　　as long as a precipitate forms, after which the
　　liquid is rendered strongly alkaline with baric
　　hydrate.  The further treatment is as described in
　　the Analysis of Potash Manures (p. 244).

*b.* **Lime and sulphuric acid.**—These are determined in two separate portions, of 50 c.c. each, of the nitric-acid-solution obtained (p. 234, *b*), the lime as described in the Analysis of Bone-dust (p. 224), the sulphuric acid with baric chloride in the usual manner.

## 6. Nitrogen—

The **ammoniacal and organic nitrogen** is determined by soda-lime-combustion of ·5 to ·8 grm. of the superphosphate.

**Ammonia.**—2 to 3 grms. are treated by Schlösing's method under a bell-jar (see p. 22); or an aqueous extract may be treated with sodic hypobromite, and the resulting nitrogen determined volumetrically in the azotometer; this method is not applicable in the presence of those nitrogenous organic bodies which, like uric acid, are decomposed similarly to ammonia by the hypobromite; nitrates, on the other hand, do not interfere with the determination.

**Nitric acid** is contained in some superphosphates to which Chili saltpetre has been added. A portion of the aqueous solution corresponding to 1 grm. of the substance is evaporated to a small volume, and the nitric acid determined by Schlösing's method (p. 116), with ferrous chloride and hydrochloric acid.

Superphosphates which have been mixed with ammonia-salts should be examined for **ammonic sulphocyanate,** which is injurious to vegetable life. Its presence is readily recognisable by the red colour imparted to the aqueous extract by a drop of ferric chloride. **Quantitatively,** the sulphocyanate is determined by treating an aqueous extract of the superphosphate with a solution of cupric sulphate to

which sulphurous acid has been added. The whitish precipitate of cuprous sulphocyanate is filtered off, washed,
dried at 115° C., and weighed as anhydrous salt $(CNSCu)_2$.

The following is given by Krocker as the composition
of a superphosphate prepared from Baker-guano :—

| | |
|---|---|
| Moisture . . . . ⎫ | |
| Chemically-combined water ⎬ | 27·00 |
| Combustible matter . . ⎭ | |
| Soluble phosphoric acid . | 21·31 |
| Insoluble ,, ,, . | 1·05 |
| Sulphuric acid . . . | 24·65 |
| Lime . . . . | 23·20 |
| Magnesia . . . . | 1·30 |
| Alkalies . . . . | ·49 |
| Insoluble matter . . | 1·00 |
| | 100·00 |

## 8. GYPSUM.

Pure crystallised gypsum consists of 46·51 % of sulphuric anhydride, 32·56 % lime, and 20·93 % water, as
represented by the formula—

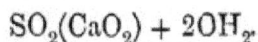

$$SO_2(CaO_2) + 2OH_2.$$

The gypsum of commerce, however, frequently contains
calcic carbonate, clay, sand, etc., as these are generally
present with the gypsum found in nature.

1. **Moisture and water of crystallisation.**—20 to 30
grms. of the air-dried gypsum are roughly pulverised, and
5 grms. are then further reduced to very fine powder in a
mortar.

3 grms. of this very finely-divided gypsum are gently ignited in a platinum crucible, and then treated with a little pure ammonic carbonate, and heated to 150° C. until the excess of the latter has been driven off. The loss in weight is calculated as moisture and water of crystallisation.

## 2. Lime, sulphuric acid, iron, alumina, etc.

*A.* By solution in hydrochloric acid—

   *a.* 1 to 1·5 grms. of the finely-divided gypsum are heated in a beaker with very dilute hydrochloric acid (to 1 grm. of substance 50 c.c. of water and 10 c.c. of strong hydrochloric acid, Sp. G. 1·12, are used). The digestion is continued until all but clay and sand has dissolved. The insoluble residue is collected on a tared filter, and, after washing well, is dried and weighed; then the filter-paper and residue are ignited and again weighed. The difference in weight is calculated as organic matter, and the final weight as clay and sand.

   *b.* The filtrate is divided into two equal parts. One half is nearly neutralised with sodic carbonate, then heated nearly to boiling, and baric chloride added to precipitate the sulphuric acid. The baric sulphate is collected on a filter, washed, dried, and ignited in the usual way; from its weight the sulphuric acid is calculated $(233 : 80 :: SO_2BaO_2$ found $: x)$, and from the latter the crystallised gypsum $(80 : 172 :: SO_3$ found $: x)$.

   *c.* The other half of the filtrate obtained in *a* is oxidised with a few drops of strong nitric acid to convert any iron (the presence of which has been indicated with potassic ferrocyanide) into ferric salt. The iron and alumina are then precipitated with ammonia; the precipitated ferric and aluminic hydrates are quickly filtered off, washed, dried, and ignited.

   *d.* The filtrate from *c* is heated, and then a solution of ammonic oxalate added. The precipitated calcic oxalate is, after subsidence, collected on a filter, well washed, dried, ignited gently, and weighed. From the weight of calcic carbonate thus obtained the percentage of lime is calculated ($100 : 56 :: COCaO_2$ found : $x$).

3. **Calcic carbonate.**—If, on treating the gypsum with hydrochloric acid, effervescence occurs, a determination of the carbonic acid should be made in the manner described in the Analysis of Marl. From the carbonic acid found the carbonate of lime is then calculated.

*B.* By decomposition with sodic carbonate—

1 grm. of the finely-divided gypsum is boiled for one hour with a solution of ten times its weight of pure sodic carbonate. In this operation the calcic sulphate is decomposed, calcic carbonate and sodic sulphate being formed. The calcic carbonate and other residue are filtered off and well washed. The filtrate is cautiously acidulated with hydrochloric acid, and the sulphuric acid precipitated by addition of baric chloride.

The residue above is treated with dilute hydrochloric acid, loss by effervescence being carefully avoided; any clay and sand remain undissolved, and in the filtrate from these the lime, etc., are determined as in *A*, *c* and *d*, above.

The following analysis of a commercial gypseous manure is given by Krocker :—

| | |
|---|---|
| Moisture and water of crystallisation . | 21·50 |
| Sulphuric acid . . . . . | 41·00 = 88·15 |
| | of crystallised |
| | gypsum. |
| Lime . . . . . . . | 28·70 |
| Calcic carbonate . . . . . | 3·50 |
| Ferric oxide and alumina . . . | 1·50 |
| Organic matter . . . . . | ·50 |
| Clay and sand . . . . . | 2·80 |
| Other substances and loss . . . | ·50 |
| | 100·00 |

## 9. NITRATE OF SODA (Chili Saltpetre).

Sodic nitrate ($NO_2NaO$) contains 63·53 % $N_2O_5$, 16·47 % N, and 36·47 % $Na_2O$ ; the commercial nitrate of soda is generally also mixed with small quantities of the sulphates and chlorides of calcium, magnesium, and sodium.

The adulteration of the nitrate with any of the above can be readily detected by ordinary **qualitative** analysis ; thus, the presence of

**Common salt** is recognised by a heavy white precipitate with argentic nitrate.

**Sodic** sulphate gives a copious white precipitate with baric chloride insoluble in hydrochloric acid.

Sodic carbonate by effervescence with hydrochloric acid.

**Magnesic and calcic sulphates,** by the precipitate with baric chloride, and by precipitation of the lime with ammonic oxalate, and subsequent precipitation of the magnesia with hydric disodic phosphate and ammonia.

The **quantitative** examination is conducted as follows:—

1. **Water.**—A considerable quantity of the nitrate is reduced to fine powder in a mortar, and 2 grms. of this finely-divided substance are dried at 100° C. until of constant weight. The loss is calculated as water.

2. **Insoluble matter.**—10 to 20 grms. of the finely-powdered nitrate are extracted with water, and the insoluble residue, consisting of sand and earthy matter, is collected on a filter, washed, dried, ignited, and weighed.

3. **Soluble matter.**—The filtrate from the above is diluted to 500 c.c. and employed for the following determinations:—

> *a.* **Chlorine.**—10 c.c. are titrated with standard argentic nitrate and potassic chromate. (See p. 51).
>
> Or 100 c.c. are acidulated with nitric acid, and the chloride precipitated with argentic nitrate and determined in the usual way.
>
> *b.* **Sulphuric acid.**—200 c.c. of the above filtrate are acidulated with hydrochloric acid, heated nearly to boiling, and then treated with baric chloride in slight excess. The precipitated baric sulphate, after complete subsidence, is collected on a filter, washed, dried, ignited, and weighed.
>
> *c.* **Lime and magnesia.**—The remainder of the above filtrate is rendered alkaline with ammonia; any precipitate is filtered off, and in the filtrate, after heating nearly to boiling, the lime is precipitated with ammonic oxalate. The calcic oxalate

R

is filtered off, dried, gently ignited, and from the weight of calcic carbonate obtained, the lime is calculated.

In the filtrate from the lime the magnesia is precipitated with hydric disodic phosphate and ammonia. After standing twelve hours the ammonic magnesic phosphate is filtered off, washed with ammonia-water (1 part strong ammonia, 3 parts water), dried, ignited, and weighed as $P_2O_3(MgO_2)_2$.

*d.* Nitric acid and sodic nitrate.—Many methods of determining this, the most important constituent of the manure, are in vogue amongst analysts, most of which are, however, extremely imperfect, and although described here cannot be recommended. The one which alone is free from objection is Schlösing's method, which has already been described.

   *a.* By loss.—After all the bases and acids, together with the water, have been determined, the loss is calculated as nitric acid.

   *β.* By ignition with silica.—The moisture and chemically-combined water, if any, are determined by gently heating 2 grms. of the finely-divided substance in a platinum crucible, and then weighing to ascertain the loss. Another 2 grms. are mixed with seven times their weight of ignited quartz-sand free from carbonates, and then ignited strongly, the operation being repeated until the weight remains constant. The loss, after subtracting the moisture, etc., determined above, is calculated as nitric acid.

γ. **By decomposition with** ferrous chloride (Schlösing).—The determination of nitric acid in sodic and potassic nitrates by this method has been already fully described.  (See p. 56.)

The following results of analysis of a commercial sample of nitrate of soda are given by Krocker :—

| Moisture | . | . | . | . | 1·20 |
|---|---|---|---|---|---|
| Sodic chloride | . | . | . | . | 1·65 |
| Calcic sulphate | . | . | . | . | ·08 |
| Magnesic nitrate | . | . | . | . | ·80 |
| Potassic nitrate | . | . | . | . | ·40 |
| Potassic sulphate | . | . | . | . | ·20 |
| Sodic nitrate | . | . | . | . | 95·47 |
| Insoluble matter | . | . | . | . | ·20 |

100·00

## 10. POTASH-MANURES.

The large deposits of potassium-salts at Stassfurth and other places in Germany have of late years become the source of most important manures, which are very extensively used in agriculture.

These manures, besides potassic chloride and potassic sulphate, generally also contain magnesic sulphate, magnesic chloride, sodic chloride, calcic sulphate, together with a small quantity of matters insoluble in water (magnesia, ferric oxide, clay, sand, etc.)

Although for judging of the agricultural value of potash-manures other than nitre it is sufficient to have a knowledge of the proportion of potassium contained, yet a complete analysis of the manure may at times be of interest, the examination being then conducted as follows :—

1. **Water.**—A considerable quantity of the material is reduced to fine powder in a mortar. 3 to 4 grms. are then ignited very gently, and the loss estimated as water.

2. **Matters soluble and insoluble in water.**—10 grms. of the finely-powdered substance are heated with about 300 c.c. of water in a flask; the insoluble residue is collected on a tared filter, washed, and dried at 100° C. The filtrate is made up to 1000 c.c., and employed for the following determinations :—

   a. **Chlorine.**—Either by titration of 10 c.c. with standard argentic nitrate and potassic chromate, or by precipitation of 50 to 100 c.c. with argentic nitrate and gravimetric determination of the precipitated argentic chloride.

   b. **Lime and magnesia.**—200 c.c. are first strongly acidulated with hydrochloric acid and then rendered alkaline with ammonia, whereupon the lime is precipitated in the hot solution with ammonic oxalate, and the magnesia in the filtrate with hydric disodic phosphate.

   c. **Sulphuric acid.**—100 to 200 c.c. are heated nearly to boiling, acidulated with hydrochloric acid, and then precipitated with baric chloride. The baric sulphate is filtered off, washed, dried, ignited, and weighed as $SO_2(BaO_2)$.

   d. **Potash**—

      a. In 50 to 100 c.c. the sulphuric acid is precipitated with baric chloride, and the liquid is then rendered strongly alkaline with baric hydrate to precipitate the magnesia. After filtration the alkaline earths are precipitated in the filtrate with ammonic carbonate and oxalate. The filtrate from the latter is weakly acidulated with hydrochloric acid, and evaporated to dryness. The dry mass is ignited to drive off ammonia-salts, and then the potassium is precipitated in the residue with platinic chloride. If the potash only is to be determined, as is generally the case, 3 grms. of the finely-powdered substance should be taken and heated with about 100 c.c. of water in a flask, the sulphuric acid is precipitated with baric chloride, and then the magnesia with baric hydrate in excess. When cold, the liquid is diluted to 250 c.c. without filtration, and 50 or 100 c.c. are measured out, filtered, and then precipitated hot with ammonic carbonate and filtered again. The filtrate is evaporated to dryness, ignited to drive off ammonia-salts, and then the potassium precipitated with platinic chloride.

β. The potash may also be determined by first precipitating the sulphuric acid with baric chloride, filtering off the baric sulphate, and then adding platinic chloride at once to the filtrate. The potassic platinic chloride is separated from the other double salts of platinum with the alkaline earths and sodium by its insolubility in alcohol, the others being soluble.

10 grms. of the finely-powdered substance are boiled with 300 c.c. of water in a flask,

and then baric chloride is added drop by drop until the whole of the sulphuric acid is precipitated, a large excess of baric chloride being avoided. When cold, the liquid is made up to 1000 c.c. without previous filtration.

100 c.c. of the clear solution (= 1 grm. of the substance), which must be filtered if necessary, are treated with a quantity of a solution of platinic chloride containing 2 grms. of platinum. The liquid is evaporated to dryness on the water-bath, and the dry residue treated with alcohol (80 °/$_o$). The insoluble potassic platinic chloride is well washed by decantation with alcohol, and collected on a tared filter; it is then dried at 100° C., and weighed as $2KCl, PtCl_4$.

Instead of 10 grms. it is generally sufficient to take 5 grms. of the substance and to boil with 100 c.c. of water. The sulphuric acid should then be exactly precipitated as above, the lamp being removed from beneath the flask each time that the baric chloride is added, and the precipitate being allowed to subside before fresh baric chloride is added. When cold, the liquid is made up to 250 c.c.; 50 c.c. (= 1 grm. substance) are filtered and used for the determination of the potassium as above.

# VIII. ANALYSIS OF MILK AND OTHER DAIRY PRODUCE.

## 1. MILK.

The composition of milk varies considerably, not only in different mammals, but also in the same species, and even in the same individual; the quality of the milk being greatly dependent upon the external circumstances to which the individual is exposed. On this account it is not always possible to ascertain by analysis whether the poverty of milk is due to artificial or natural causes.

The following table shows the average composition, expressed in parts per 100, of the milk of some of the more common mammals:—

|  | Cow. | Goat. | Ewe. | Mare. | Ass. | Bitch. | Sow. | Woman. |
|---|---|---|---|---|---|---|---|---|
| Density . . | 1·0318 | 1·0323 | 1·038 | 1·031 | 1·033 | 1·036 | 1·044 | 1·0315 |
| Solid residue | 13·50 | 12·40 | 18·00 | 11·00 | 9·30 | 26·30 | 23·00 | 12·30 |
| Caseine . . | 3·60 | 3·70 | 6·10 | 2·70 | 1·70 | 11·70 | 12·89 | 1·90 |
| Butter . . | 4·05 | 4·20 | 5·33 | 2·50 | 1·55 | 9·72 | 6·60 | 4·50 |
| Sugar . . | 5·50 | 4·00 | 4·20 | 5·50 | 5·80 | 3·00 | ·50 | 5·30 |
| Extractive Matters and Salts | ·40 | ·56 | ·70 | ·50 | ·50 | 1·35 | 3·01 | ·18 |
| Water. . . | 86·45 | 87·54 | 83·67 | 88·80 | 90·45 | 74·23 | 77·00 | 88·12 |
|  | 100·00 | 100·00 | 100·00 | 100·00 | 100·00 | 100·00 | 100·00 | 100·00 |

According to Grandeau, the composition of cow's milk may vary from natural causes within the following limits:

| | | | | | | |
|---|---|---|---|---|---|---|
| Water | . | . | . | . | 80·00 to 88·65 |
| Fat | . | . | . | . | 2·90 „ 4·50 |
| Caseine | . | . | . | . | 3·00 „ 5·00 |
| Albumen | . | . | . | . | ·30 „ ·55 |
| Milk-sugar | . | . | . | . | 3·00 „ 5·50 |
| Ash | . | . | . | . | ·70 „ ·80 |

The complete analysis of milk is conducted as follows :—

1. Total solids and water.—The evaporation of milk is attended with no little difficulty, owing to the fine skin which forms upon the surface when the milk is heated. The formation of this skin is prevented by previously coagulating the caseine with acetic acid.

The evaporation of the milk, and subsequent dessication of the residue, is also accelerated by the addition of some indifferent but anhydrous powder, such as sand, glass, or baric sulphate.

If the solid residue is to be subsequently used for other determinations, such as fat, etc., then 50 grms. of milk are introduced into a tared glass dish containing about 10 grms. of baric sulphate and a little glass rod, the whole having been dried and weighed at 100° C. After introducing the 50 grms. of milk a few drops of acetic acid are added, and the dish is placed upon a water-bath, the milk being stirred during the evaporation with the little glass rod. The residue is then dried in an air-bath at 100° to 105° C., and weighed.

If the solid residue is not to be further utilised, then it is sufficient to evaporate 4 grms. of milk as above; in fact it is generally advisable to determine the water and total solids by a special evaporation of 4 grms. of milk, as the thorough dessication of the residue from a larger quantity is exceedingly troublesome and tedious.

2. **Fat (Butter)**—

    *a.* The residue obtained from the 50 grms. of milk in (1) is reduced to a fine powder, and washed into a little flask with ether. The flask is fitted with an inverted Liebig's condenser, and heated on a water-bath. The ethereal extract is filtered into a small flask, and the extraction of the residue is repeated several times with fresh quantities of ether. The extraction with ether may also be performed by means of the apparatus described on p. 181. The whole ethereal extract is evaporated in a weighed glass dish, and dried at 100° C. The increase in the weight of the dish gives the fat or butter in 50 grms. of milk.

    *b.* **Optical method of Vogel.**—Although in all cases in which strict accuracy is necessary the above process must be adopted, yet often, when the quality of various samples of milk is merely to be compared, this optical method may, on account of its rapidity, be with advantage substituted.

    The principle of the process is, that when milk is added to a given quantity of water contained in a vessel with glass sides (5 m.m. apart), the liquid will the sooner become opaque to the light of a standard candle the richer the milk is in butter. The relation between the amount of milk which must be added and the percentage of butter in the milk has been determined by experiment.

The requisites for the process are :—

  a. A measuring cylinder graduated to 100 c.c.
  b. A cell with parallel glass sides, the inner surfaces
of which are exactly $\frac{1}{2}$ centimetre apart.
  c. A graduated pipette (5 c.c.), graduated to $\frac{1}{4}$ c.c.
  d. A case with two slits in which to place the cell
whilst the observations are made, so as to exclude
light from the side.
  e. A standard candle.

The measuring cylinder is filled up to 100 c.c.
with distilled water ; about 3 c.c. of the well-mixed
milk are run into the cylinder from the pipette ;
the open end of the cylinder is closed with the
hand, and inverted so that the milk and water
become thoroughly mingled. The cell is then filled
with the mixture, placed in its case, and the candle
observed through the slit.

In making the observation the candle is placed
at a distance of 1 to 2 feet from the cell, the exact
distance being immaterial. It is very important,
however, that the eye of the observer should be
approximated as closely as possible to the cell. The
background behind the candle should be dark, the
window being at the back of the observer. All
side-light, whether from the window or from the
candle, must be carefully excluded.

If the candle be still visible, the liquid is poured
back from the cell into the measuring cylinder, and
$\frac{1}{2}$ c.c. of milk is added from the pipette, and after
mixing with the water as before the transparency
of the mixture is again tested. The addition of
milk, $\frac{1}{4}$ to $\frac{1}{2}$ c.c. at a time, is continued until the

contour of the candle-flame is no longer distinguish-
able, and even the halo of light from the same disappears. (In the case of ordinary cow's milk about
3 c.c. of milk are required to render the water
opaque.)

The percentage of butter can then be calculated
from the following table :—

| c.c. of Milk used. | Butter, % by Weight. | c.c. of Milk used. | Butter, % by Weight. | c.c. of Milk used. | Butter, % by Weight. |
|---|---|---|---|---|---|
| 2·50 | 9·51 | 5·00 | 4·87 | 7·50 | 3·32 |
| 2·75 | 8·73 | 5·25 | 4·66 | 7·75 | 3·22 |
| 3·00 | 7·96 | 5·50 | 4·45 | 8·00 | 3·13 |
| 3·25 | 7·41 | 5·75 | 4·26 | 8·25 | 3·04 |
| 3·50 | 6·86 | 6·00 | 4·09 | 8·50 | 2·96 |
| 3·75 | 6·44 | 6·25 | 3·94 | 8·70 | 2·88 |
| 4·00 | 6·03 | 6·50 | 3·80 | 9·00 | 2·80 |
| 4·25 | 5·70 | 6·75 | 3·66 | 9·25 | 2·73 |
| 4·50 | 5·38 | 7·00 | 3·54 | 9·50 | 2·67 |
| 4·75 | 5·13 | 7·25 | 3·43 | 9·75 | 2·61 |

The percentage of fat in the milk may also be calculated
by means of the formula—

$$\text{Fat }\% = \frac{23\cdot2}{y} + 0\cdot23,$$

in which $y$ = c.c. of milk used in the optical test.

It is frequently of importance to ascertain the quantity
of butter produced per day. For this purpose the milk
obtained at each milking should be well mixed and accurately measured in litres, and then the percentage of butter

determined by the above method in a fair sample of the milking. By also taking the specific gravity of the milk, its weight may be determined, and taking $500$ grms $= 1$ lb., the weight of butter in lbs. is easily calculated. Thus, if the Sp. G. of the milk be $1·030$, then 1 litre weighs $1030$ grms., and $\dfrac{°/_{\circ} \text{ of butter} \times 1030}{100} =$ grms. of butter in 1 litre of milk.

### 3. Milk-sugar, caseine, etc.—

a. The residue insoluble in ether from $(2, a)$ above, is collected on a filter, dried at $100°$ C., and weighed. It is then repeatedly heated with alcohol ($80$ °/$_{\circ}$), which dissolves the milk-sugar; the residue insoluble in alcohol is collected on a filter, dried at $100°$ C., and weighed. The loss in weight is due to milk-sugar; the residue consists of caseine and insoluble salts. The residue is incinerated, the ash treated with ammonic carbonate, gently ignited, and weighed. After subtracting the weight of baric sulphate used in the original evaporation, the difference gives the insoluble salts.

b. The milk-sugar can be directly estimated with greater accuracy as follows :—

20 grms. of milk are diluted with about 30 c.c. of water, and heated to $50°$ C., with a few drops of acetic acid to coagulate the caseine. The curds are collected on a filter of fine linen, washed, and the filtrate made up to 250 c.c. The latter is then digested for one hour with a few drops of sulphuric acid on the water-bath, in order to convert the milk-sugar (lactose) into an equal weight of

glucose. The liquid is then neutralised with caustic soda, and the glucose determined with Fehling's solution in the usual way.

4. **Albumenoids** (caseine, etc.) are directly determined by evaporating 6 grms. of milk with 1 grm. of ignited baric sulphate or gypsum and a few drops of acetic acid on the water-bath. In the dry residue the organic nitrogen is determined by combustion with soda-lime in the usual way. The albumenoids are then calculated by multiplying the nitrogen found by the factor 6·25.

5. **Ash.**—30 grms. of milk, after curdling with a few drops of acetic acid, are evaporated to dryness in a platinum crucible on the water-bath, and the residue carbonised. The carbonaceous mass is repeatedly boiled with water, and the aqueous extracts are filtered and evaporated to dryness in a platinum dish. The carbon and filter-paper are incinerated, and the ash added to the contents of the platinum dish; the latter is then treated with ammonic carbonate, gently ignited, and weighed.

## 2. CREAM.

Since the analysis of cream is conducted in a precisely similar manner to the analysis of milk, it will be sufficient to quote the results of analysis of cream made by Völcker and others.

The following are four analyses made by Völcker[1] of cream obtained from pure milk after standing different lengths of time :—

[1] On "Milk," *Journ. Roy. Agric. Soc.* **xxiv.** 1863.

|              |   I.    |   II.   |  III.  |   IV.   |
|--------------|---------|---------|--------|---------|
| Water . .    |  74·46  |  64·80  |  56·50 |  61·67  |
| Fat . . .    |  18·18  |  25·40  |  31·57 |  33·43  |
| Caseine . .  |  2·69 } |         |        | { 2·62  |
| Milk-Sugar . |  4·08 } |  7·61 } |  8·44  | { 1·56  |
| Ash . . .    |  0·59   |  2·19   |  3·49  |  0·72   |
|              | 100·00  | 100·00  | 100·00 | 100·00  |

| Nitrogen .   | ·43 %/₀ |   ...   |  ...   |  ·42 %/₀ |

Specific Gravity
  at 62° F   1·0194   1·0127      ...        ...

No. I.  Cream was obtained from milk after standing fifteen hours.

No. II.  After the milk had stood forty-eight hours. This is to be regarded as a type of good cream.

Nos. III. and IV.  After the milk had stood forty-eight hours; both are very rich in fat.

The following average composition of cream on the one hand, and of skimmed milk on the other, is due to Alex. Müller of Stockholm :—

|                    |      | Milk. | Skimmed Milk. | Cream.        |
|--------------------|------|-------|---------------|---------------|
| Water . .          | %/₀  | 86·81 | 89·60         | 52·0 to 63.   |
| Caseine and Sugar  | „    | 8·47  | 8·49          | 6·3 „ 7·6.    |
| Fat . . .          | „    | 3·97  | 1·19          | 40·6 „ 39·3.  |
| Ash . . .          | „    | ·75   | ·80           | ·42.          |

## 3. BUTTER

The composition of butter is subject to great variations, according to the manner in which it is made. The best butter not only contains the largest proportion of fat, but also has less tendency to become rancid than inferior butter containing more water and caseine. The following composition of several kinds of butter given by Grandeau may be regarded as typical:—

| | English | | Brunswick. | Stockholm. | Lorraine. | Vosges. |
|---|---|---|---|---|---|---|
| | 1. | 2. | 3. | 4. | 5. | 6. |
| Fat | 79·72 | 82·70 | 80·70 | 90·18 | 85 | 83 |
| Caseine. | 3·38 | 2·45 | 2·80 | 1·87 | 4 | 5 |
| Water . | 16·90 | 14·85 | 13·50 | 6·10 | 10 | 11 |
| Salt | ... | ... | 3·00 | 1·85 | 1 | 1 |
| | 100·00 | 100·00 | 100·00 | 100·00 | 100 | 100 |

The inferior kinds of butter in all countries are frequently mixed with fat from other sources. Thus, much of the butter exported from Switzerland is mixed with lard, tallow, etc.; whilst a large quantity of Dutch butter imported into this country is adulterated with "Oleomargarine," obtained by melting beef-fat at 95° F., at which temperature the greater part of the oleine and margarine of the fat is rendered fluid, whilst the stearine remains behind in the solid state.

The analysis of butter is conducted as follows:—

1. **Water.**—About 20 grms. of butter are introduced into a tared test-tube, which is then again weighed. The

butter is now dried on a water-bath, and finally in an air-bath at 100° C. until of constant weight.   The loss is calculated as water.

2. **Fat.**—The residue obtained in 1 is extracted with ether in the manner described in the Analysis of Milk.  The ethereal extract is passed through a tared hot filter, the insoluble residue being washed with ether until free from fat.  The filtrate containing the fat is evaporated to dryness in a tared glass dish on the water-bath and weighed.

3. **Caseine and ash.**—The residue insoluble in ether from 2 is washed on the filter with water, dried, and weighed ; after which it is incinerated, treated with a little ammonic carbonate, ignited gently, and the weight of the ash determined.   By subtracting the ash from the previous weight the **caseine** is obtained.

4. **Sugar and salt.**—The aqueous washings in 3 contain both the sugar and the salt.   The solution is made up to a definite volume, and in one portion the milk-sugar is estimated by first converting into glucose and then titrating with Fehling's solution (see Analysis of Milk), whilst in the other portion the **common salt** is determined either by evaporation or by titration with standard argentic nitrate and potassic chromate.

# 4. CHEESE.

The composition of cheese is also subject to great variation, according to its mode of preparation.   Thus the caseine varies from 15 to 40 °/$_o$, the fat from 20 to 40 °/$_o$, whilst the remainder consists of water, salt, milk-sugar, and the mineral constituents of the milk, which are for the most part removed from the latter along with the curds.

1. **Water, fat, and caseine.**—The cheese is cut up into small cubes, a definite weight (2 to 5 grms.) of which is taken and dried in a dessicator over sulphuric acid. The cubes are then introduced into a small tared flask, crushed with a glass rod, and then covered with about 30 c.c. of ether. The flask is now corked and allowed to stand for several days, being frequently shaken during this time. The ethereal extract is poured into a tared flask and the ether distilled off; after drying at 100° C. the flask is weighed, the increase being **fat**. After the whole of the fat has been removed by repeated digestion as above from the cheese, the flask is dried at 100° C. and weighed. By subtracting the fat and the residue insoluble in ether from the weight of cheese taken, the weight of **water** in the cheese is obtained; whilst by washing the residue insoluble in ether with water, weighing again, and subtracting the weight of ash found as below, the weight of **caseine** is obtained.

2. **Nitrogen and ash.**—The residue insoluble in ether and water obtained in 1 is mixed with soda-lime, and employed for combustion in the usual way.

Or the residue may be used for the determination of the ash. A separate portion (2 to 3 grms.) of the cheese may also be incinerated in a large crucible. The ash is treated with a little ammonic carbonate ignited gently and weighed.

Some old cheeses sometimes contain **ammonia**, which can be estimated in the usual way, by mixing the cheese with caustic potash and allowing it to stand for about ten days under a bell-jar with a standard solution of sulphuric acid.

3. **Sugar.**—The difference in weight between the original cheese employed and the sum of the fat, caseine,

S

and ash, represents, in young cheeses at least, the weight
of the milk-sugar.

The sugar may also be directly determined in the
aqueous washings from the caseine in 1, by titration with
Fehling's solution (see Analysis of Milk).

# IX. THE ANALYSIS OF WATER.

THE analysis of water has of late years become so important, that in the following pages this subject will be dealt with somewhat more fully than might at first appear warrantable.

The services which water analysis can render to agriculture are of the greatest importance ; for by frequent analysis of the drainage-water flowing from his fields the farmer is enabled to ascertain whether or not his soil is being impoverished by the method of drainage he has employed, whilst by the analysis of liquid manures he is able to estimate the agricultural value of any system of irrigation.

The accuracy with which the analysis of waters can be performed is to a great extent dependent upon the use of apparatus for the analysis of gases which is frequently not accessible to the chemist. But since the analysis of waters in connection with agriculture can generally be accomplished satisfactorily without such apparatus, the subject will be divided into—

1. WATER ANALYSIS WITHOUT GAS-APPARATUS.

2. WATER ANALYSIS WITH GAS-APPARATUS.

# 1. WATER ANALYSIS WITHOUT GAS-APPARATUS.

## 1. COLLECTION OF SAMPLES.

It is of the greatest importance that samples of water for analysis should be collected with the utmost care, and not left to incompetent persons.

The water should be collected in stoppered bottles, " Winchester " quarts being the most convenient for the purpose. The bottles must be perfectly clean, and it is preferable that they should not have been previously used for any other purpose. Stoneware jars should if possible be avoided, as, even if clean, the clay, which frequently contains calcic sulphate, is liable to affect the hardness of the water. Corks are also objectionable, but if used they must be quite new and have been well washed with the water.

Before collecting the sample the bottle should be three times well rinsed with the water; it is then filled to within half an inch of the stopper, and the latter tied down tightly with a piece of string or tape.

If the water is taken from a pump, about 4 gallons, and if from a tap about 2 gallons, should be allowed to flow before collecting, and the water should then be made to pass directly from the spout into the bottle.

When the sample is intended to represent the water-supply of a town, it should be taken from a pipe in direct communication with a street-main and not from a cistern.

In taking a sample of water from a tank, well, or stream, the mouth of the bottle should be completely immersed below the surface of the water if practicable; and if a can or dipper be used, both it and the string used in drawing it up must be scrupulously clean, and the collection of surface water avoided as much as possible. On the other hand, care must be taken not to disturb any mud or sediment at the bottom of the water.

The following quantities of water should be collected for ordinary analysis :—

Sewage.  
Polluted rivers and streams. } 1 Winchester quart.  
Shallow-well waters.

Deep wells.  
Ordinary unpolluted rivers, } 2 Winchester quarts.  
   streams, and springs.

Lakes and tarns. } 3 Winchester quarts.  
Mountain springs and streams.

When a complete mineral analysis of the solid residue is required, at least double the above quantities should be collected.

In collecting the samples the following information should also be obtained concerning the water :—

   *a.* The **source,** whether from **well, river,** or **stream.**

      If from **well,**—

   *b.* Description of the soil and subsoil, also the water-bearing stratum into which the well is sunk.

   *c.* The diameter and depth of well.

   *d.* Distance of the well from either cesspools or drains.

      If from **river or stream,**—

   *e.* Distance from the source to the point at which the sample is collected.

   *f.* Whether sewage or other animal polluting matter

is known to gain access to the river or stream
above the point at which the sample is collected.

If from **spring**,—

g. The stratum from which the spring issues.

h. Whether the sample is taken direct from the
spring or otherwise.

## 2. Taste, Odour, Colour, and Turbidity.

The **taste and odour** of water are most marked when
the water is made lukewarm.

The **colour** is determined by viewing a stratum of the
water about 2 feet in thickness against a white ground.
This may be done by filling a glass cylinder of that height
with the water and standing it upon a white tile.  It is
well to compare the colour with that of a standard water
of known purity placed in a similar cylinder.

The **turbidity** of the water is observed by shaking up
the sample and pouring out about $\frac{1}{4}$ litre into a flask, and
then holding the latter up to the light, a portion of the
field of view being darkened by some object behind.

The analyst must now decide whether the water is
to be filtered or not before further examination.  His
decision must depend upon the manner in which the water
is used.  If the amount of suspended matter be inconsider-
able, and would naturally be carried with the water in its
movements, then filtration is unnecessary, but the sample
should always be shaken before withdrawing a portion for
analysis.  On the other hand, if the quantity of suspended
matter is large, a measured volume of the well-shaken
water ($\frac{1}{2}$ litre or $\frac{1}{4}$ litre) should be filtered through a
tared filter (dried at 100° C.); the filter-paper and residue,

after being dried at 100° C., are weighed, the increase in weight being the **total suspended matter.** The residue and filter are then incinerated in a platinum crucible, and the ash, after being moistened with a solution of ammonic carbonate, is dried at 150° C. and weighed. The weight of the residue, after subtracting that of the filter-ash, gives the **mineral suspended matter,** the difference between which and the total represents the organic suspended matter.

### 3. Total Solid Matter in Solution.

$\frac{1}{4}$ or $\frac{1}{2}$ litre of the filtered or unfiltered water, as the case may be, is evaporated in a platinum dish that has been ignited to redness and tared after cooling in a dessicator. The evaporation is carried on upon a water or steam-bath, the dish being supported upon a smooth glass ring. When all the water is evaporated, the dish containing the residue is placed in an air-bath at 100° C. for three hours. The dish is then allowed to cool in a dessicator and weighed; it is now placed in the air-bath and weighed at intervals of half-an-hour until of constant weight. The weight of the residue is calculated to 1 litre of water, and this result must then be multiplied by 100 to be expressed in parts per 100,000, or by 70 to be expressed in grains per gallon.

It is sometimes necessary not only to determine the amount of the total solid matter in solution, but also to make a complete analysis of the mineral ingredients of this solid matter. Such information is frequently required by brewers in order to ascertain the fitness of water for brewing particular liquors.

### 4. COMPLETE MINERAL ANALYSIS OF SOLID RESIDUE.

The complete mineral analysis of the residue is conducted according to methods which have been already fully described, but which may be here again briefly summarised.

Silica.—Not less than 1 litre of the water is evaporated to dryness in a platinum dish with a few drops of hydrochloric acid. The residue is ignited until no more acid fumes are evolved; when cool it is repeatedly extracted with hot dilute hydrochloric acid, and finally with hot water; the insoluble silica is collected on a filter.

Iron and alumina.—The filtrate from the silica is treated with a few drops of strong nitric acid, boiled, and a slight excess of ammonia added. After the precipitate has subsided the clear liquid is passed through a filter; the precipitate is dissolved in a little hydrochloric acid again precipitated with ammonia, and collected on the filter, which, after washing, drying, and igniting, is weighed as ferric and aluminic oxides.

Lime.—The filtrate from the iron and alumina is treated with ammonic oxalate in excess. The precipitate is collected on a filter, and, after washing and drying, it is ignited strongly over the blowpipe and weighed as calcic oxide.

Magnesia.—The filtrate from the lime is concentrated by evaporation and then treated with ammonia and hydric disodic phosphate. The precipitated ammonic magnesic phosphate is collected on a filter and weighed as magnesic pyrophosphate, $P_2O_3(MgO_2)_2$.

Sulphuric acid.—1 litre of the water is acidulated with hydrochloric acid, and after reducing the bulk to about 100 c.c. an excess of baric chloride is added, and the precipitated **baric sulphate** collected on a filter and weighed.

**Phosphoric acid.**—1 litre of the water is acidulated with nitric acid, and then evaporated down to a volume of about 50 c.c., which is treated with a solution of ammonic **molybdate.** The precipitate, if there be any, is treated in the usual way.

**Alkalies.**—1 litre of the water is treated with a few drops of baric chloride to precipitate the sulphuric acid; some pure baric hydrate is then added to precipitate iron, alumina, magnesia, and phosphoric acid. After boiling, the liquid is filtered, and the filtrate, after concentration by evaporation, is treated with ammonia, ammonic carbonate, and ammonic oxalate; the liquid is again filtered, and the filtrate evaporated to dryness and ignited gently until free from ammonia-salts. The residue is extracted with hot water, filtered if necessary, acidulated with hydrochloric acid and evaporated to dryness in a tared platinum dish, and the mixed **alkaline chlorides,** after gentle ignition, weighed. The potassium is then precipitated by treating the residue with platinic chloride, evaporating to dryness on the water-bath, and extracting the sodic chloride and excess of platinic chloride with alcohol. The insoluble potassic platinic chloride $(2KCl, PtCl_4)$ is collected on a tared filter, washed with alcohol, dried at 100° C., and weighed.

## 5. AMMONIA.

Owing to the liability of this ingredient in water to change, its estimation should be proceeded with without delay.

The ammonia is estimated by the intensity of the

colouration produced with Nessler's **Reagent**, by which
1 part of ammonia in 100,000,000 of water can be recognised.

The quantity of water used for the determination will
vary according to the quality of the water.   Half a litre is
usually a suitable quantity, but 1 litre may be taken of
deep-well and river-water, whilst ¼ litre or less of shallow-
well water will generally be found sufficient.

The water taken, whatever its volume, is distilled in a
stoppered retort of about 3 litres capacity, and attached to
a Liebig's condenser.   The retort is first two-thirds filled
with ordinary water, to which a few grms. of ignited sodic
carbonate have been added.   The apparatus is freed from
ammonia by distilling off about 100 c.c. of this water until
the distillate ceases to give any colouration with the Nessler
solution.   The water under examination is now introduced
by means of a funnel through the tubulure of the retort,
and the distillation proceeded with, the retort being heated
with the free flame of a large bunsen-burner.   The distillate
is collected in colourless glass cylinders of about 50 c.c.
capacity.   The whole of the ammonia, unless an excessive
amount be present, is contained in the first three cylinders
of the distillate.

The three cylinders are placed upon a white tile and
covered with watch-glasses, 1 c.c. of Nessler's solution being
added to each of the cylinders.

The ammonia in each cylinder is now estimated by com-
paring the colour with that produced by the Nessler in a
similar cylinder filled with distilled water free from ammonia
(see below), and to which a known volume of a standard
solution of ammonic chloride (see below) has been added.
A number of test-cylinders are prepared in this way until

the exact tint of each of the three cylinders containing the distillate has been imitated.

The colouration produced when more than 2 c.c. of the standard solution of ammonic chloride are added to a cylinder of distilled water is too intense for accurate comparison. Hence, when the second cylinder of the distillate is found to contain a quantity of ammonia greater than that corresponding to ·5 c.c. of the standard solution of ammonic chloride, the first cylinder will always contain an amount of ammonia which is too great to be estimated as above. The contents of the first cylinder should then be diluted with distilled water free from ammonia, so as to fill two or more cylinders, and the ammonia in each of these then estimated as above.

The quantities of ammonia found in each of the distillates are then added up, and the total calculated to parts per 100,000.

If the water contain urea, the evolution of ammonia in the distillation will be long continued. In such cases the distillate should be collected in smaller quantities, and only the ammonia which is rapidly evolved at first should be estimated, the remainder being neglected.

The estimation of ammonia in water is of great importance, as the presence of this compound in any but very minute quantities is generally indicative of recent pollution with animal organic matters. The amount of ammonia in sewage usually varies from 2 to 10 parts per 100,000, and in waters polluted by sewage between even still wider limits; thus in shallow-well water it varies from 0 to 2·75. The presence of ammonia in deep-well water is less suspicious, as it is there often derived from the reduction of nitrates. Thus many deep-well waters of excellent quality, obtained from the Lower London Tertiaries under the London Clay, contain a considerable quantity of ammonia derived from this origin.

**Preparation of Nessler's solution.**—62·5 grms. of potassic iodide are dissolved in about 250 c.c. of distilled water ; about 10 c.c. of this solution are set aside, and then a cold saturated solution of mercuric chloride is added to the remainder until a permanent precipitate is formed. This precipitate is now redissolved by the reserve of potassic iodide, and then mercuric chloride is added very cautiously until a slight precipitate remains on stirring. 150 grms. of caustic potash dissolved in distilled water are now added, and the whole made up to 1 litre. After subsidence, the clear yellow liquid is decanted off and kept in a stoppered bottle ready for use.

**Standard solution of ammonic chloride.**—1·5735 grm. of pure dry ammonic chloride are dissolved in 1 litre of water, and 100 c.c. of this strong solution are diluted to 1 litre for use. 1 c.c. of the dilute solution contains ·00005 grm. $NH_3$.

**Sodic carbonate free from ammonia.** — The dry carbonate is heated strongly for one hour in a platinum dish without fusing. It should be preserved in a small stoppered bottle ready for use.

**Water free from ammonia.**—Distilled water, or ordinary good water, is distilled with sodic carbonate, the distillate being rejected as long as it produces a colouration with the Nessler. The distillation must not be pushed too far, and the distillate coming over at the end of the operation should be again tested with the Nessler.

## 6. CHLORINE.

50 c.c. of water are measured out into a small flask standing upon a white tile, a few drops of a neutral solution of potassic chromate are added, and then a standard solution of argentic nitrate is run in from a burette until the liquid has acquired a faintly-red tinge. The completion of the reaction is sometimes a little difficult to recognise owing to the turbidity produced by the precipitation of the argentic chloride, and it is well in such cases to have as a standard of comparison a second flask, to which an insufficient quantity of argentic nitrate has been added.

The chlorine in water is principally combined as common salt; and this ingredient, although perfectly harmless in itself, is nevertheless objectionable when present in large quantity, as it is generally indicative of the water having been polluted with the liquid excrements of animals. Thus human urine contains about 500 parts of chlorine, or 824 parts of sodic chloride, in 100,000 parts. The Rivers Pollution Commissioners found from their numerous analyses the average proportion of chlorine per 100,000 parts was, for

| | | |
|---|---|---|
| Rain-water . . . . . | ·22 | part. |
| Unpolluted upland-surface water . | 1·13 | ,, |
| ,,  deep-well water . . | 5·11 | ,, |
| ,,  spring-water . . | 2·49 | ,, |

**Standard argentic-nitrate-solution.**—2·3944 grms. of argentic nitrate are dissolved in 1 litre of distilled water. If 50 c.c. of the water are employed, then the number of c.c. of silver solution used indicate at once the parts of chlorine per 100,000.

Potassic-chromate-solution.—A strong solution of
the pure neutral potassic chromate, free from chloride, is
employed.

## 7. HARDNESS.

The hardness, or soap-destroying power of water is caused
by the presence of lime and magnesia salts in solution,
and the hardness is expressed as equivalent to so many
parts of calcic carbonate dissolved in 100,000 parts of
the water.

The determination of the hardness is carried out as
follows :—

Into a stoppered bottle of about 8 oz. capacity 50 c.c. of
the water are introduced, and well agitated by shaking ;
the stopper is then withdrawn, and the air sucked out by
means of a tube. In this way any carbonic acid that may
be given off from the water on shaking is removed.
Standard soap-solution (see below) is now added from
a burette, 1 c.c. at a time, the bottle being vigorously
shaken after each addition. The addition of soap-solution
is continued more cautiously until the lather, which forms
upon the surface of the water after shaking and laying the
bottle on its side, remains unbroken for five minutes. The
point at which the reaction is completed can also be recognised by the ear when the bottle is shaken, the sound
becoming much smoother and softer as the end of the
operation is reached. From the number of c.c. used, and
the table below, the degree of hardness of the water can be
at once determined. If more than 16 c.c. of soap-solution
are required, the operation must be repeated upon a smaller
volume (25 or 10 c.c.) of water, the volume being made
up to 50 c.c. with recently-boiled distilled water before

adding the soap-solution. The degrees of hardness found must then be multiplied according to the degree of dilution.

When magnesia-salts are present the curd has a characteristic lightness, and a permanent lather sometimes appears long before the operation is really completed, the lather being dissipated on repeating the shaking. In such cases the water should be diluted so that less than 7 c.c. of soap-solution are required for 50 c.c. of the diluted water.

Temporary hardness.—Most waters become softer after boiling, owing to the decomposition of the soluble bicarbonates of lime and magnesia; carbonic acid being set free, and the insoluble neutral calcic carbonate being deposited—

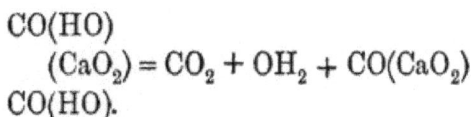

$$CO(HO) \atop (CaO_2) = CO_2 + OH_2 + CO(CaO_2) \atop CO(HO).$$

This hardness, which is due to the bicarbonates of lime and magnesia, and which disappears on boiling the water, is known as **temporary** hardness, whilst that which remains after boiling as permanent hardness. The temporary and permanent hardness are determined as follows :—

A flask of about 300 c.c. capacity, of which the neck has been cracked off short, is made up to a weight of 200 grms. with the water. The water is then boiled briskly for half an hour, the steam escaping freely from the short neck; the weight is then again made up to 200 grms. with distilled water, and when cold 50 c.c. are taken and titrated with soap-solution as above. The hardness so found is **permanent**, and by subtracting it from the total hardness the **temporary** is obtained.

## TABLE OF HARDNESS IN PARTS PER 100,000, 50 C.C. OF WATER BEING USED.

| c.c. of Soap-Solution. | CaCO₃ per 100,000. | c.c. of Soap-Solution. | CaCO₃ per 100,000. | c.c. of Soap-Solution. | CaCO₃ per 100,000. | c.c. of Soap-Solution. | CaCO₃ per 100,000. | c.c. of Soap-Solution. | CaCO₃ per 100,000. |
|---|---|---|---|---|---|---|---|---|---|
| ·7 | ·00 | 3·8 | 4·29 | 6·9 | 8·71 | 10·0 | 13·31 | 13·1 | 18·17 |
| ·8 | ·16 | ·9 | ·43 | 7·0 | ·86 | ·1 | ·46 | ·2 | ·33 |
| ·9 | ·32 | 4·0 | ·57 | ·1 | 9·00 | ·2 | ·61 | ·3 | ·49 |
| 1·0 | ·48 | ·1 | ·71 | ·2 | ·14 | ·3 | ·76 | ·4 | ·65 |
| ·1 | ·63 | ·2 | ·86 | ·3 | ·29 | ·4 | ·91 | ·5 | ·81 |
| ·2 | ·79 | ·3 | 5·00 | ·4 | ·43 | ·5 | 14·06 | ·6 | ·97 |
| ·3 | ·95 | ·4 | ·14 | ·5 | ·57 | ·6 | ·21 | ·7 | 19·13 |
| ·4 | 1·11 | ·5 | ·29 | ·6 | ·71 | ·7 | ·37 | ·8 | ·29 |
| ·5 | ·27 | ·6 | ·43 | ·7 | ·86 | ·8 | ·52 | ·9 | ·44 |
| ·6 | ·43 | ·7 | ·57 | ·8 | 10·00 | ·9 | ·68 | 14·0 | ·60 |
| ·7 | ·56 | ·8 | ·71 | ·9 | ·15 | 11·0 | ·84 | ·1 | ·76 |
| ·8 | ·69 | ·9 | ·86 | 8·0 | ·30 | ·1 | 15·00 | ·2 | ·92 |
| ·9 | ·82 | 5·0 | 6·00 | ·1 | ·45 | ·2 | ·16 | ·3 | 20·08 |
| 2·0 | ·95 | ·1 | ·14 | ·2 | ·60 | ·3 | ·32 | ·4 | ·24 |
| ·1 | 2·08 | ·2 | ·29 | ·3 | ·75 | ·4 | ·48 | ·5 | ·40 |
| ·2 | ·21 | ·3 | ·43 | ·4 | ·90 | ·5 | ·63 | ·6 | ·56 |
| ·3 | ·34 | ·4 | ·57 | ·5 | 11·05 | ·6 | ·79 | ·7 | ·71 |
| ·4 | ·47 | ·5 | ·71 | ·6 | ·20 | ·7 | ·95 | ·8 | ·87 |
| ·5 | ·60 | ·6 | ·86 | ·7 | ·35 | ·8 | 16·11 | ·9 | 21·03 |
| ·6 | ·73 | ·7 | 7·00 | ·8 | ·50 | ·9 | ·27 | 15·0 | ·19 |
| ·7 | ·86 | ·8 | ·14 | ·9 | ·65 | 12·0 | ·43 | ·1 | ·35 |
| ·8 | ·99 | ·9 | ·29 | 9·0 | ·80 | ·1 | ·59 | ·2 | ·51 |
| ·9 | 3·12 | 6·0 | ·43 | ·1 | ·95 | ·2 | ·75 | ·3 | ·68 |
| 3·0 | ·25 | ·1 | ·57 | ·2 | 12·11 | ·3 | ·90 | ·4 | ·85 |
| ·1 | ·38 | ·2 | ·71 | ·3 | ·26 | ·4 | 17·06 | ·5 | 22·02 |
| ·2 | ·51 | ·3 | ·86 | ·4 | ·41 | ·5 | ·22 | ·6 | ·18 |
| ·3 | ·64 | ·4 | 8·00 | ·5 | ·56 | ·6 | ·38 | ·7 | ·35 |
| ·4 | ·77 | ·5 | ·14 | ·6 | ·71 | ·7 | ·54 | ·8 | ·52 |
| ·5 | ·90 | ·6 | ·29 | ·7 | ·86 | ·8 | ·70 | ·9 | ·69 |
| ·6 | 4·03 | ·7 | ·43 | ·8 | 13·01 | ·9 | ·86 | 16·0 | ·86 |
| ·7 | ·16 | ·8 | ·57 | ·9 | ·16 | 13·0 | 18·02 | ... | ... |

Preparation of standard calcic-chloride-solution. —·2 grm. of pure calcic carbonate is carefully dissolved in

a platinum dish with dilute hydrochloric acid.   The excess
of acid is driven off by repeatedly evaporating to dryness
with distilled water; the residue, when free from acid, is
dissolved in water and made up to 1 litre.

**Standard soap-solution.**—40 parts of dry potassic
carbonate and 150 parts of lead-plaster (*Emplastrum plumbi*,
B.P.) are rubbed together in a mortar until well mixed.
Methylated spirit is now added, and the mixture triturated
to a cream.   After standing for some hours it is transferred
to a filter, and repeatedly washed with methylated spirit.
The strength of the filtrate is ascertained by running it
from a burette into 50 c.c. of the standard calcic-chloride-
solution, as in the determination of hardness.   The solution is then diluted with methylated spirit and water until
exactly 14·25 c.c. are required to produce a permanent
lather with the 50 c.c. of the standard calcic chloride.   In
diluting the soap-solution as above, water must be added
so that in the whole the volume of water to that of methylated spirit is as one to two.   The soap-solution should
be made too strong at first, and after allowing to stand
for twenty-four hours it is filtered if necessary, and then
accurately diluted.

## 8. Nitrogen as Nitrates and Nitrites.

With the exception of the small proportion contained in
rain-water, and due to the combination of atmospheric
nitrogen and oxygen during the electrical discharge, the
whole of the nitric and nitrous acids present in water are
derived from the oxidation of animal organic matter.

Of the processes already described for the determination

T

of nitric acid, only Warington's modification of Schlösing's method is applicable to the analysis of waters.

The numerous methods which have been devised for the determination of nitric nitrogen in water are all based upon one of three principles :—

    *a.* The decomposition of nitric and nitrous acids into nitric oxide, and the gasometric measurement of the latter.

    *b.* The reduction of nitric and nitrous acids to ammonia by nascent hydrogen, and subsequent estimation of the ammonia formed.

    *c.* The bleaching action of nitric acid upon a standard solution of indigo.

    *a.* Of the two methods belonging to this category that of Schlösing has already been fully described. The other, known as the Crum-Frankland method, will be discussed in the second part of Water Analysis, inasmuch as it involves the use of gas-apparatus.

    *b.* Zinc-copper method—

The nitrogen, as nitrates and nitrites, is reduced by means of an electrical couple of precipitated copper and zinc (for preparation of, see below). The residue obtained in the determination of the total solid matter is treated with about 25 c.c. of distilled water, and then boiled down to one-fourth of its volume with a fragment of recently-ignited lime about the size of a hemp-seed, in order to destroy any urea that may be present. The liquid is then transferred to a small flask, and the dish containing the residue is rinsed out with distilled water free from ammonia until the volume of liquid in the flask is about 15 to 20 c.c.

A Wurtz-flask (8 oz. capacity) fitted to a small Liebig's condenser is well adapted for effecting the reduction and subsequent distillation in. The requisite quantity (about 5 grms.) of zinc-copper couple is placed in the flask, the mouth is fitted with a cork bearing a stoppered funnel, and the distillation is then commenced. The distillate is collected in the same cylinders (50 c.c.) used in the determination of ammonia. When almost all the liquid has passed over, hot distilled water is admitted into the flask through the funnel, and the distillation continued until about two cylinders of distillate have been collected.

Nearly the whole of the ammonia is contained in the first cylinder, and it is generally necessary to dilute this before estimating the amount with Nessler's solution, whilst the second cylinder may be treated with the Nessler at once.

If used at once, the apparatus may be employed for a second sample of water without adding a fresh quantity of zinc-copper couple.

**Preparation of zinc-copper couple.**—Thin sheet-zinc cut into small pieces is placed in a flask and covered with a tolerably-concentrated solution of cupric sulphate. The action is allowed to continue for ten to fifteen minutes, and then the supernatant liquid is poured off, and the flask filled up several times with cold distilled water to wash the precipitated copper. The operation may be conducted on a small scale (3 grms. of zinc) in the flask in which the reduction and distillation is subsequently to be performed, and the couple will then be ready for immediate use.

c. **Aluminium method**—

The reduction of the nitric and nitrous acids is in this

method effected by the nascent hydrogen produced in the action of aluminium-foil on caustic soda or potash :—

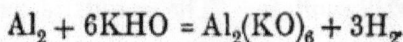

$$Al_2 + 6KHO = Al_2(KO)_6 + 3H_2$$

100 c.c. of water are placed in a small flask, together with 10 c.c. of a solution of caustic soda (10 %) free from nitrates, and the liquid reduced by ebullition to one-fourth its bulk. The original volume is then restored with distilled water free from ammonia, and when cold a piece of aluminium-foil about 2 in. square, and wrapped round a piece of glass rod to ensure its sinking, is introduced. The evolved hydrogen is made to pass through a small U-tube containing powdered glass moistened with hydrochloric acid (free from ammonia) to arrest any ammonia that may be carried over ; a second tube is attached to this to protect it from any atmospheric ammonia.

The reducing action is allowed to continue for about six hours, and then the contents of the first U-tube are transferred to the flask, a small Liebig's condenser is attached, and the liquid distilled until the distillate passes over free from ammonia. The ammonia in the distillate is estimated by means of Nessler's solution as above.

### d. Indigo method—

This method — which has been recently improved by Warington — when carefully carried out according to the following instructions, yields very trustworthy results :—

" 4 grms. of sublimed indigotine are digested for some hours with five times their weight of Nordhausen sulphuric acid ; the liquid is then diluted with water, filtered, and brought to the volume of 2 litres. The strength of the indigo is next ascertained with a normal solution of nitre by the method presently to be described. The indigo-solution, which will be found too strong, is finally diluted, so that 10 c.c. correspond with the same volume of the normal nitre-solution. There is no object in bringing the indigo absolutely to the normal strength—a near approximation will suffice. To preserve the indigo-solution Sutton recommends that 4 °/₀ by volume of pure oil of vitriol should be added to it; indigo so fortified undergoes no change when kept in the dark.

"The normal indigo-solution is diluted, as required, to four times its volume for the purpose of water analysis, oil of vitriol being added in such quantities as to maintain the proportion of 4 per cent. The indigo-solution is to be used in a burette graduated to lengths of a cubic centimetre. As the solution is too dark to be seen through, the height of the liquid may be read from its upper surface, or a float may be employed. The solution being slightly viscid, the burette should not be read till the level of the liquid becomes constant.

"A series of standard solutions of pure nitre will be required to determine the value of the indigo in nitric acid, as this is not uniform throughout the scale. A normal solution of nitre is first prepared by dissolving 1·011 grm. of pure nitre (gently fused over a spirit-lamp before weighing) in 1 litre of water. From this normal solution, other solutions of $\frac{1}{4}$, $\frac{1}{8}$, $\frac{1}{16}$, $\frac{1}{32}$, and $\frac{1}{64}$ the normal strength are prepared, and preserved for use.

"In water analysis it is most convenient to work with 20 c.c. of water; when this is done it is necessary to standardise the indigo with 20 c.c. of the $\frac{1}{8}$, $\frac{1}{16}$, $\frac{1}{32}$, and $\frac{1}{64}$ nitre-solutions. If 10 c.c. of water are to be employed, the indigo must be also standardised with 10 c.c. of the $\frac{1}{4}$ and $\frac{1}{8}$ nitre-solutions. It is most useful to standardise both for 10 c.c. and 20 c.c. of water, as waters of a greater range of strength can then be analysed without dilution. It is unnecessary to standardise with 10 c.c. of the weaker nitre-solutions; if the amount of nitrate in a water is not greater than that in $\frac{1}{8}$ nitre-solution, 20 c.c. of the water may be at once taken for the estimation.

"A large supply of **pure** distilled **oil of** vitriol will be required for extensive water analysis. It should be colourless, quite free from nitrous compounds, and contain as little as possible of sulphurous acid, and must be of nearly full gravity. As any variation in the acid affects the determination, it is best to mix the contents of all the bottles purchased before proceeding to standardise the indigo. The oil of vitriol is measured for use in a tolerably wide burette, provided with a glass stopcock.

"A further requisite is a **chloride of calcium** bath, provided with a thermometer; the bath is conveniently made in a porcelain basin. The temperature to be maintained is 140° C.; and as the temperature keeps rising from the evaporation of the solution, it is necessary to bring it down to the required point by the addition of a little water or chloride of calcium solution immediately before each experiment. The chloride of calcium bath is not required when strong solutions of nitrate are analysed by a normal solution of indigo, the reaction in such cases being almost immediate. With weak solutions of indigo and nitrate, the reaction may take some time,—in extreme cases as much as five minutes,—and it becomes essential for accuracy that the temperature should be maintained throughout at the normal point.

"The standardising of the indigo-solution is performed as follows:—

"In a wide-mouthed flask of about 150 c.c. capacity are placed 10 c.c. or 20 c.c. of the standard nitre-solution; as much indigo-solution is measured in as is judged sufficient, and the whole mixed. Oil of vitriol is next run from the burette into a test-tube, in quantity exactly equal to the united volumes of the nitre and indigo-solutions. The contents of the test-tube are then poured as suddenly as possible into the solution in the flask, the whole rapidly mixed, and the flask at once transferred to the chloride of

calcium bath.  It is essential for concordant results that
the oil of vitriol should be uniformly mixed with the
solution as quickly as possible.  This is especially necessary
in the case of strong solutions of nitrate, in which the
action begins immediately after the addition of the
sulphuric acid ; with such solutions it is more difficult to
get duplicate experiments to agree than with weaker
solutions, when the action does not begin at once, and in
which, therefore, time is afforded for mixing.  The operator
should not attempt to drain the test-tube ; the oil of vitriol
adhering to the tube is a fairly constant quantity, and after
the first experiment the tube will deliver the quantity
measured into it.

"It is well to have the flask covered by a watch-glass
while holding it in the chloride of calcium bath.  The
progress of the reaction should be watched, and as soon as
the greater part of the indigo has been oxidised, the contents of the flask should be gently rotated for a moment.
With very weak solutions of pure nitre, no change is
observable for some time, and it may be necessary in some
cases to keep the flask in the chloride of calcium bath for
five minutes.  If the colour of the indigo is suddenly
discharged, it is a sign that the nitric acid is in considerable excess, and that a much larger amount of indigo must
be taken for the next experiment.  If some of the indigo
remains unoxidised, a little experience will enable the
operator to judge its probable amount, and so decide on
the quantity of indigo suitable for the next experiment.

"The amount of indigo which corresponds to the solution of nitrate is found by a series of approximating
experiments made, as just described, with varying quantities
of indigo, the oil of vitriol used being always equal in
volume to the united volumes of the nitrate-solution and
indigo.  The determination is finished when a quantity of
indigo is left unoxidised not exceeding ·1 c.c. of the indigo-
solution used ; this amount can be readily estimated by
the eye.  It is well, until considerable experience has been

gained, to check the result by making a further experiment
with ·1 c.c. less indigo, when the colour should be entirely
discharged.   The tint produced by a small excess of indigo
is best seen by filling up the flask with water.   The
estimated excess of indigo is, of course, deducted from
the reading of the burette.

"To reduce the number of experiments required to obtain the result, it is well to proceed with some boldness, and
ascertain as soon as possible what are the limits between
which the quantity of indigo must fall.   Seven experiments will be a maximum rarely exceeded, and four about
the average required, where the character of the waters is
already known.

"The direction has been given to employ oil of vitriol
equal in volume to the united nitrate and indigo-solutions.
It is most important that the same proportion of oil of
vitriol should be used in every experiment, as the quantity
of indigo oxidised is much affected by the proportion of
acid present, being considerably greater when the acid
equals the volume of the other solutions than when the
acid is used in larger proportion.

"When the indigo-solution has been standardised with
the series of nitre-solutions already mentioned, it will be
found that the quantity of indigo consumed is not strictly
in proportion to the nitric acid present, but diminishes
as the nitrate-solution becomes more dilute.   In round
numbers, a diminution of the amount of nitre present
to $\frac{1}{8}$ is accompanied by diminution of the indigo oxidised
to $\frac{1}{10}$, or, in other words, if 20 c.c. of the $\frac{1}{8}$ normal
solution of nitre require 10 c.c. of indigo, 20 c.c. of the
$\frac{1}{64}$ nitre-solution will require only 1 c.c. of indigo.   This is
a very important fact, and necessitates the standardising of
the indigo with solutions of graduated strength, so that the
value of the indigo may be known for all parts of its scale.

"Experience has shown that, with a $\frac{1}{2}$ normal-solution
of indigo, 10 c.c. of a $\frac{1}{4}$ normal-solution of nitre will oxidise
distinctly less than half the indigo required by 10 c.c. of a

$\frac{1}{2}$ normal nitre-solution, and with the $\frac{1}{4}$ normal solution of indigo employed in water analysis the differences produced by dilution are still more considerable.

"It becomes necessary, then, to form a table of the value in nitrogen corresponding to each part of the indigo scale, and by the help of this table every analysis subsequently made is calculated. Below is given an ideal table of this description. It is assumed, which is very near the truth, that a diminution to one-eighth in the strength of the nitrate-solution is accompanied by a diminution to one-tenth in the indigo consumed. It is further assumed, which is also near the truth, that the alteration in the relation of the indigo to the nitrate proceeds at a uniform rate between the limits actually determined. The following will be the results arrived at when using 20 c.c. of the nitrate-solution for each experiment :—

## "VALUE OF INDIGO, IN NITROGEN, FOR DIFFERENT STRENGTHS OF NITRE-SOLUTIONS.

| Strength of Nitre-solution used. | Indigo required. | Difference between amounts of Indigo. | Nitrogen corresponding to 1 c.c. of Indigo. | Difference between the Nitrogen Values. | Difference in the Nitrogen Values for a difference of 1 c.c. in the amount of Indigo. |
|---|---|---|---|---|---|
| | | c.c. | grm. | grm. | grm. |
| $\frac{8}{64}$ normal | 10·00 c.c. | ... | ·000035000 | ... | ... |
| $\frac{7}{64}$ " | 8·71 " | 1·29 | ·000035161 | ·000000161 | ·000000125 |
| $\frac{6}{64}$ " | 7·43 " | 1·28 | ·000035330 | ·000000169 | ·000000132 |
| $\frac{5}{64}$ " | 6·14 " | 1·29 | ·000035627 | ·000000298 | ·000000231 |
| $\frac{4}{64}$ " | 4·86 " | 1·28 | ·000036008 | ·000000381 | ·000000298 |
| $\frac{3}{64}$ " | 3·57 " | 1·29 | ·000036764 | ·000000756 | ·000000586 |
| $\frac{2}{64}$ " | 2·29 " | 1·28 | ·000038209 | ·000001445 | ·000001129 |
| $\frac{1}{64}$ " | 1·00 " | 1·29 | ·000043750 | ·000005541 | ·000004295 |

"The mode of using this table is very simple. Supposing that 20 c.c. of a water have required 3·5 c.c. of indigo, this amount is seen to be ·5 c.c. above the nearest point (4·86 c.c.) given in the table. We learn from the right-hand column that ·000000149 must consequently be subtracted from the unit value in nitrogen (·000036008 grm.) belonging to 4·86 c.c. of indigo. We thus find that the 5·36 c.c. of indigo should be reckoned at ·000035859 grm. of nitrogen per c.c. The water, therefore, contains ·96 part of nitrogen as nitric acid per 100,000. If 20 c.c. of the water have required less than 1 c.c. of indigo, the unit value corresponding to 1 c.c. of indigo is employed for calculating the amount; this will give an accurate result if the oil of vitriol used be quite pure.

" It is not necessary in practice to form so complete a table as that now given; it will suffice to determine the indigo with nitre-solutions of $\frac{1}{8}$, $\frac{1}{16}$, $\frac{1}{32}$, and $\frac{1}{64}$ normal strength; if another solution is used, that of $\frac{3}{64}$ normal strength seems the most important. It may be sufficient even in some cases simply to determine the indigo with the $\frac{1}{8}$ and $\frac{1}{24}$ nitre-solutions, and to find all the intermediate values by calculation. If 10 c.c. of water are to be employed as well as 20 c.c., separate tables should be formed, giving the value of the indigo in each case.

" There is one important advantage that follows from the troublesome necessity of standardising the indigo with various nitrate-solutions: it tends to remove the errors due to impurities in the oil of vitriol. As the volume of the oil of vitriol used depends far more on the volume of water taken than on the quantity of nitric acid it contains, all errors due to oxidising or reducing matter contained in the oil of vitriol fall far more heavily on determinations of small quantities of nitric acid than on determinations of larger quantities. Unless, therefore, the oil of vitriol used was absolutely pure, it would always be necessary to standardise the indigo as now recommended, even if the relation of indigo to nitric acid were exactly the same in weak and strong solutions. By standardising as now described, the errors due to the sulphuric acid affect the figures of the table, but not the result of the analysis. Such standardising has also the effect, as far as it goes, of

calibrating the burette, any inequalities of capacity falling
on the table, and not on the analysis calculated from it.

"In standardising the indigo-solution, and in the subsequent experiments with it, some regard must be paid to
the initial temperature of the solutions. A rise in the
initial temperature will be attended by a diminution in the
quantity of indigo oxidised; this is most perceptible in
the case of the stronger solutions of nitrate. The following table shows the results obtained by standardising the
same indigo-solution at two temperatures, representing
nearly the extreme limits at which the solution would
be used in practice. The temperature of the room during
the experiment was 10°. For the trials at the higher temperature, the flask containing the indigo and nitrate, and the
test-tube containing the oil of vitriol, were placed for some
time in a water-bath at 22° to 23° previously to being mixed.

## "INDIGO-SOLUTION STANDARDISED WITH 10 C.C. OF NITRE-SOLUTION.

| Strength of Nitre-Solution. | At 10° C. | | At 22° C. | |
|---|---|---|---|---|
| | Indigo required (actual). | Indigo calculated from two extremes. | Indigo required (actual). | Indigo calculated from two extremes. |
| | c.c. | c.c. | c.c. | c. c. |
| $\frac{1}{4}$ normal . . | 10·28 | 10·28 | 9·76 | 9·76 |
| $\frac{1}{8}$ ,, . . | 4·97 | 4·97 | 4·78 | 4·73 |
| $\frac{1}{16}$ ,, . . | 2·30 | 2·32 | 2·18 | 2·22 |
| $\frac{1}{32}$ ,, . . | 0·99[1] | 0·99 | 0·96 | 0·96 |
| STANDARDISED WITH 20 C.C. OF NITRE-SOLUTION. | | | | |
| $\frac{1}{8}$ normal . . | 10·26 | 10·26 | 9·74 | 9·74 |
| $\frac{1}{16}$ ,, . . | 4·84 | 4·96 | 4·65 | 4·74 |
| $\frac{1}{32}$ ,, . . | 2·21 | 2·31 | 2·21 | ·24 |
| $\frac{1}{64}$ ,, . . | 0·99 | 0·99 | 0·99 | 0·99 |

[1] "It must not be supposed that it was possible to work to $\frac{1}{100}$th of
a cubic centimetre. In all the figures given the errors shown by cali-

"It is seen that a rise of 12° in temperature diminishes
the indigo consumed by about 5 per cent in the case of
the stronger solutions of nitrate.  It is evident, therefore,
that the indigo-solution should be standardised at nearly
the same temperature at which it is to be used.

"In the table the quantity of indigo corresponding to
the intermediate strengths of nitre-solution has been calculated from the determinations made with the two
extreme strengths, to show the amount of accuracy to be
attained by this plan.  The agreement of calculation with
the results actually obtained is sometimes very close, but
in others the difference is somewhat greater than the errors
of experiment.  If the indigo-solution has been standardised with 20 c.c. of the nitre-solutions recommended, the
operator will be able to analyse waters containing nitric
acid up to 1·75 part of nitrogen per 100,000.  If the
indigo has also been standardised with 10 c.c. of the $\frac{1}{4}$ and
$\frac{1}{8}$ nitre-solutions, the operator can, by employing 10 c.c. of
the water, extend the range of his analysis to waters containing 3·4 parts of nitrogen per 100,000.  Waters stronger
than this should be diluted for analysis.  The natural error
of the determination, shown by the performance of duplicate experiments, will not exceed 1 per cent of the nitric
acid present in the case of waters containing 1·7 of nitrogen
per 100,000, but may amount to 5 per cent with waters
containing as little as 0·2 of nitrogen per 100,000.

"Chlorides are, of course, generally present in waters containing nitrates; they are not altogether without effect on
the determination with indigo, but the error thus introduced is not sufficient to be of importance.

"The general conclusion respecting the indigo method for
determining nitric acid will be that it is excellently adapted
by its simplicity, rapidity, and delicacy, for general use in
water analysis; but that accuracy can be secured only by

brating the measuring vessels employed have been taken account of,
and minute fractions thus introduced."

working under the same conditions which obtained when the indigo-solution was standardised. In the presence of organic matter, the indications obtained with indigo must be accepted as probably below the truth."

## 9. NITRITES.

The only reliable method for the determination of nitrites in water is that devised by Griess, and depending upon the yellow colouration produced when a solution of meta-phenylene-diamine is added to nitrous acid. The reaction which takes place may be expressed as follows :—

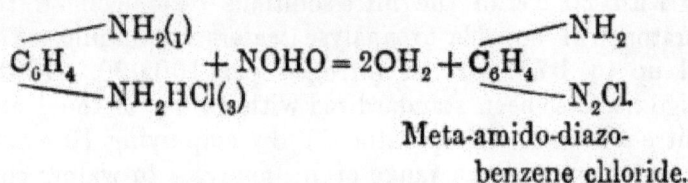

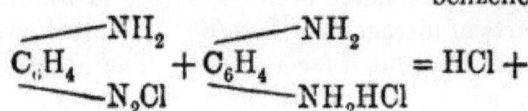

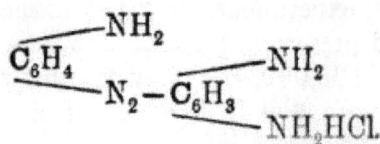

$$C_6H_4{\overset{-NH_2(_1)}{\underset{-NH_2HCl(_3)}{}}} + NOHO = 2OH_2 + C_6H_4{\overset{-NH_2}{\underset{-N_2Cl.}{}}}$$

Meta-amido-diazo-<br>benzene chloride.

$$C_6H_4{\overset{-NH_2}{\underset{-N_2Cl}{}}} + C_6H_4{\overset{-NH_2}{\underset{-NH_2HCl}{}}} = HCl +$$

$$C_6H_4{\overset{-NH_2}{\underset{-N_2-C_6H_3{\overset{-NH_2}{\underset{-NH_2HCl}{}}}}{}}}$$

Triamido-azobenzene-hydro-<br>chlorate<br>(Yellow colouring matter).

So delicate is this method, that 1 part of nitrous acid in 30,000,000 parts of water may still be detected.

100 c.c. of the water are placed in a glass cylinder, and 1 c.c. of dilute sulphuric acid (1 part of acid to 2 parts of distilled water), and 1 c.c. of a solution of metaphenylene-

diamine (for preparation see below) added. Should a red colouration be at once produced on stirring, showing that too much nitrite is present for estimation, the operation must be repeated on 50 c.c. or less of the water, diluting to 100 c.c. with water free from nitrites. The colour ought only to make its appearance after the liquid has stood for one or two minutes.

The colour produced is now imitated by treating a standard solution of potassic nitrite (for preparation see below) in a precisely similar manner. The water and the trial solutions must be treated with the reagent simultaneously, as the colouration increases on standing.

Coloured waters must be first decolourised by treating with a few drops of caustic and carbonate of soda, or by means of a little alum, and then rendering alkaline. The water must be filtered before being used for the test.

**Preparation of metaphenylene-diamine-solution.** —5 grms. of the base are dissolved in 1 litre of water, and after decolourising with animal charcoal, if necessary, the solution is acidulated with sulphuric acid.

**Preparation of standard solution of potassic nitrite.**—·406 grm. of pure dry argentic nitrite is dissolved in hot water and decomposed with a slight excess of a solution of potassic chloride. When cool the solution is made up to 1 litre, the argentic chloride is allowed to settle, and each 100 c.c. of the clear supernatant liquid is further diluted to 1 litre. This solution will contain ·00001 grm. of $N_2O_3$ in every c.c.; it should be kept in a bottle full up to the stopper.

## 10. POISONOUS METALS.

The presence of poisonous metals such as lead and

copper, which form insoluble and dark-coloured sulphides, is readily detected by acidulating 50 c.c. of the water placed in a glass cylinder with a few drops of acetic acid, and then adding a drop of ammonic sulphide. A dark colouration indicates the presence of poisonous metals, which should then be further tested for as follows :—

**Lead.**—100 c.c. of the water are acidulated with a few drops of acetic acid, and 5 c.c. of a saturated solution of sulphuretted hydrogen are then added. The colouration produced is imitated by treating a standard solution of lead in a precisely similar manner. (The standard solution of lead is made by dissolving ·1831 grm. of crystallised normal plumbic acetate in 1 litre of distilled water, then 1 c.c. = ·0001 grm. of metallic lead.)

**Copper** is estimated in just the same way as the lead, a standard solution of copper being used for comparison instead. (The standard solution of copper is made by dissolving ·3929 grm. of crystallised cupric sulphate in 1 litre of distilled water, then 1 c.c. = ·0001 grm. of metallic copper.)

If both lead and copper be present, it is necessary to evaporate a large volume of the water to dryness, then to separate the two metals from the residue with sulphuretted hydrogen in a dilute hydrochloric-acid-solution, after which the two sulphides are dissolved in dilute nitric acid, the solution evaporated to dryness with sulphuric acid, and ignited until all the nitric acid has been expelled. The residue is extracted with water acidulated with sulphuric acid, and the insoluble **plumbic sulphate** collected on a filter and weighed. The **copper** is precipitated in the filtrate with sodic hydrate or sodic hyposulphite.

**Arsenic,** which is sometimes met with in waters that

have been polluted with the drainage from mines and manu-factures, is best estimated by **Marsh's test**.

A $\frac{1}{2}$ litre of the water, after being rendered alkaline with caustic soda or potash free from arsenic, is evaporated to dryness. The residue is extracted with strong hydrochloric acid, and the extract introduced through a funnel into a small flask, in which hydrogen is being generated from pure zinc (free from arsenic). The escaping gas is made to pass through a tube containing pumice moistened with plumbic acetate, and then through a piece of combustion-tube drawn out at the end to a jet, and about 6 in. long and $\frac{1}{8}$ in. in diam. The middle of this tube is heated to redness by means of a lamp, by which any arseniuretted hydrogen is decomposed and the arsenic deposited in the cooler part of the tube. The operation should be continued for one hour, although the greater part of the arsenic comes over in the first ten minutes.

The quantity of arsenic present is determined by com-paring the deposit with a number of similar deposits obtained in the same way with known quantities of arsenic.

**Barium** may be detected by acidulating some of the water after concentrating, filtering off any turbidity, and then adding a solution of calcic sulphate.

**Zinc.**—When present in waters it is generally as bicar-bonate, which at the surface becomes converted into the normal carbonate forming a film on the water. This film should be collected and heated on platinum-foil, when, if zinc be present, a residue will be left which is yellow when hot and white when cold.

## 11. Organic Matter in Solution.

The organic matter in solution is the determination

U

which offers most difficulty in water analysis. No process
yet devised deals with the determination of more than two
of the elements (carbon and nitrogen) composing this organic
matter; whilst the only process which will be described in
this section of water analysis merely compares the amount
of organic matter contained in different waters by the
oxygen which they respectively abstract from a solution of
potassic permanganate. This method is known as the "For-
chammer" or oxygen process. In the subsequent part
of water analysis the accurate determination of the organic
carbon and organic nitrogen in water by the Frankland or
combustion process will be fully described.

## FORCHAMMER OR OXYGEN PROCESS.

As already stated, this process depends upon the varying amounts of oxygen absorbed from a solution of potassic
permanganate by the same volume of waters of different
organic purity. The process is carried out as follows :—

Two flasks are washed first with strong sulphuric acid
and then with water. 250 c.c. of the water are measured
into one flask, whilst an equal volume of distilled water, free
from organic matter (see below), is placed in the other. 10
c.c. of dilute sulphuric acid (1:3) and 10 c.c. of a standard
solution of potassic permanganate (see below) are added to
each, and the mixture allowed to stand for three hours.

The excess of permanganate is now estimated by means of
potassic iodide and a standard solution of sodic hyposulphite.
For this purpose a few drops of a solution of potassic iodide
(10 °/₀) are added to each of the flasks. A quantity of
iodine is now set free corresponding in amount to the excess
of permanganate-solution added. The quantity of this free
iodine is now determined by titrating with a standard solu-

tion of sodic hyposulphite (see below), a drop of a clear solution of starch being added towards the end of the re-action, the hyposulphite being added until the blue colour produced by the starch is just discharged.

The same operation is then performed with the flask containing the distilled water, and the amount of hypo-sulphite required is noted in each case.

Let the hyposulphite required in the case of the distilled water be denoted by A, and that in the case of the water under analysis by B, then the permanganate consumed during the reaction is represented by $A - B$. If, further, the oxygen available in the permanganate added be $a$, then the oxygen consumed by the water in question is

$$\frac{(A - B)a}{A.}$$

Thus, if 10 c.c. of permanganate, containing ·001 grm. of available oxygen, be added to $\frac{1}{4}$ litre of distilled water, and also to the same volume of the water under examination, and if the former afterwards require 40 c.c. and the latter 15 c.c. of the hyposulphite, then the oxygen consumed by the $\frac{1}{4}$ litre of water is

$$\frac{(40 - 15)\ ·001}{40} \text{ or } \frac{(40 - 15)\ ·4}{40}$$

= ·250 part of oxygen per 100,000 parts of water.

The estimation of organic matter by the above process is vitiated by the presence of other reducing substances, such as nitrites, ferrous salts, sulphuretted hydrogen, etc. Moreover, the quantity of permanganate reduced by differ-ent kinds of organic matter is very variable, so that it is only when the organic matter is of the same kind that comparable results can be obtained by this method.

**Standard solution of potassic permanganate.**—·395 grm. of the crystallised salt are dissolved in 1 litre of distilled water; 10 c.c. are equivalent to ·001 grm. of available oxygen. The solution requires restandardisation at frequent intervals.

**Standard solution of sodio hyposulphite.**—1 grm. of the salt dissolved in 1 litre of water. The solution loses its strength so rapidly that it requires restandardisation by means of a blank experiment before each set of determinations undertaken.

**Preparation of distilled water free from organic matter and ammonia.**—The blank experiment described above must not be made with ordinary distilled water, which always contains a very appreciable proportion of organic matter, but with water which has been carefully freed from organic matter and ammonia by the following process:—

About 15 to 20 litres of ordinary distilled water are treated with 1 grm. of caustic potash and ·2 grm. of potassic permanganate per litre, and digested for twenty-four hours at about 100° C., in a still with the condenser inverted. The water is then distilled off, and the distillate tested from time to time with Nessler's solution, none being collected until 100 c.c. show no colouration whatever. The distillation must not be pushed too far, and the last portion collected should be again carefully tested.

If the distilled water so obtained is to be employed in the determination of organic elements by the combustion process (to be described in the second part), it must be acidulated with a very little sulphuric acid and redistilled.

**Starch solution.**—1 part of starch is rubbed with 20 parts of boiling water, the liquid is filtered, boiled, and, after standing twenty-four hours, the clear portion siphoned off.

## 2. WATER ANALYSIS WITH GAS-APPARATUS.

As already mentioned, the most satisfactory method of determining the organic elements in water is by submitting the residue left on evaporation to combustion with oxide of copper. The quantities of carbonic anhydride and nitrogen obtained in the operation are usually so small that only gasometric methods are applicable to their accurate measurement.

Besides, in the "**Combustion Process**" of determining the organic elements, the gasometric method is also employed in the so-called "**Mercury Process**" for the determination of nitrogen as nitrates and nitrites.

These two processes will be described in the following pages.

### THE COMBUSTION PROCESS FOR DETERMINATION OF ORGANIC CARBON AND ORGANIC NITROGEN.

The process may be divided into four heads—

1. The evaporation of the water.
2. The preparation of the residue for combustion.
3. The combustion.
4. The volumetric measurement of the gases obtained.

### 1. The evaporation—

As soon as the ammonia in the water has been determined, the evaporation should be commenced. The quantity of water used for the purpose should vary inversely as its purity, but never exceeds 1 litre. Half a litre is generally sufficient, whilst if the water contain more than ·2 of ammonia in 100,000, $\frac{1}{4}$ litre, and if more than 1·0 part of ammonia, 100 c.c. of the water will suffice for evaporation.

The water is measured into a flask and there treated with 20 c.c. of a saturated solution of **sulphurous** acid (for preparation see below), after which it is boiled briskly for a few seconds.  By this means the **carbonates** are decomposed with evolution of carbonic anhydride, the **nitrates and nitrites** with evoluton of free nitrogen; whilst the **ammonia** is fixed as ammonic sulphite, and the nitrogen corresponding to it, minus a small quantity lost during evaporation (see below), must afterwards be subtracted from the nitrogen obtained by combustion.

The evaporation of the water, after treatment as above, is conducted in a glass dish about 4 inches in diameter hemispherical in shape, and without a lip, which is heated by means of a water-bath arranged as in Fig. 11.

*a a* is a flanged copper capsule, into which some ordinary water is poured, and upon this the glass dish floats. *b* is a self-filling water-bath upon which the copper-capsule rests.  The flange of the copper-capsule has a rim running round it, except at one point where it is turned down to form a lip.  Upon the circumference of the copper-capsule rests the copper-ring (*c c*), 3 inches high, and somewhat conical in shape, and bearing upon its upper rim a thin glass shade which thus covers the glass dish above.  The water to be evaporated is placed in the flask (*e*), the ground neck of which fits accurately into the self-regulating supply-tube (*f f*), whereby the water is only permitted to flow into the dish when the junction (*g*) is above the water-level.  The flask (*e*) is supported upon the wooden ring of a large filter-stand, out of which a small segment has been cut to allow of the passage of the neck.  The supply-tube rests in a notch cut in the copper-ring (*c*), just deep enough to admit it without touching the glass shade, so as to prevent the water condensing on the inside of the latter from running down into the dish.

During evaporation the water condensed on the inner surface of the glass shade runs down the latter and collects in the

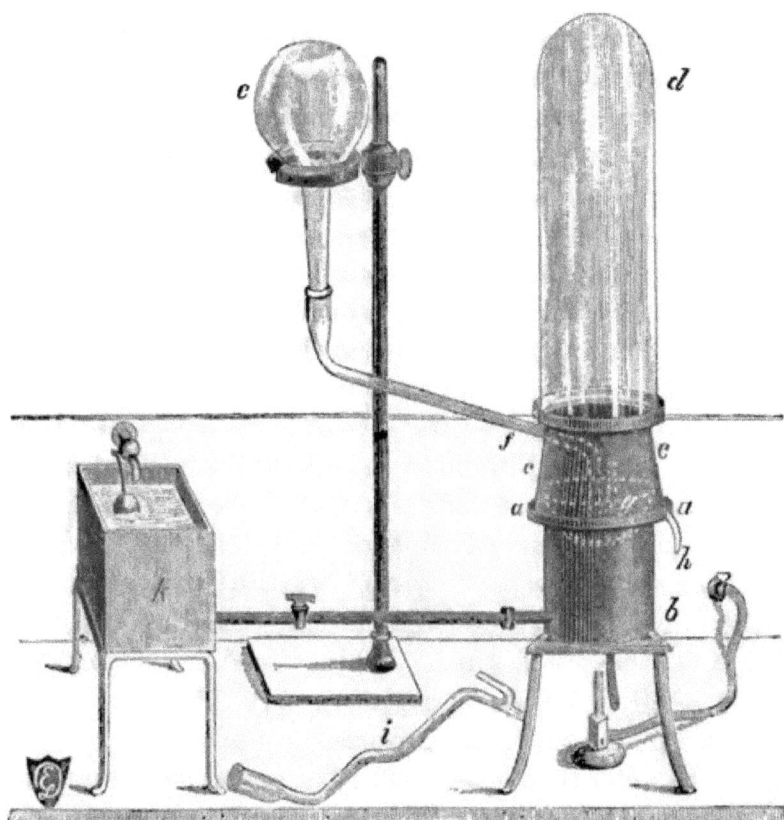

Fig. 11.

copper-capsule beneath the glass-dish, and from here it is led away by means of a piece of tape ($h$) placed over the lip.

The water must not be introduced into the flask ($e$) until quite cool, otherwise the glass joint at $f$ may afterwards become loose. After fitting on the supply-pipe ($ff$)

the flask is rapidly inverted, the open end being held over
the dish to prevent loss.   The flask is rested on the filter-
ring, and the glass shade put on the conical copper-ring,
and the evaporation commenced by lighting the lamp
beneath the water-bath.

As soon as the residue in the dish is dry (which in the
case of a litre of water will be in about twenty-four hours)
the latter is removed, and placed under a glass shade until
the combustion can be commenced.

The combustion of the residue cannot be with safety
commenced until the nitrates have been determined, as, if
the water contains a large amount of nitrates, the 20 c.c.
of sulphurous acid will be insufficient for their decomposition, and the residue will have to be treated with further
quantities.   Thus, if the nitrogen, as nitrates and nitrites,
exceed ·5 per 100,000 parts of water, the dish containing
the residue from 1 litre must be filled with distilled water
free from organic matter (see p. 292), to which $\frac{1}{10}$ its
volume of the saturated sulphurous-acid-solution has been
added, and the evaporation again carried to dryness.   If
the nitrogen, as nitrates and nitrites, exceed 1 part per
100,000, the residue from 1 litre should be evaporated
with $\frac{1}{4}$ litre, if 2 parts with $\frac{1}{2}$ litre, and if 5 parts 1 litre
of this 10 °/₀ solution of sulphurous-acid-solution.

If the water contain an insufficient proportion of bases to
combine with the free sulphuric acid formed by the oxidation
of the sulphurous acid, a small quantity (1 or 2 c.c.) of a
saturated solution of hydric sodic sulphite should be added.

Waters containing much ammonia,[1] but no nitrates, are

_____
[1] The organic nitrogen in strongly ammoniacal waters containing
no nitrates is determined with greater accuracy by evaporating with
*boracic acid* or *borax* instead of sulphurous or phosphoric acids.  The

more advantageously evaporated with 10 c.c. of a solution of **metaphosphoric acid** (for preparation see below) instead of the sulphurous acid, as ammonic phosphate loses less ammonia during evaporation than ammonic sulphite. The residue is in this case always viscid, and should be dried by the addition of about ·5 grm. of calcic phosphate (for preparation see below).

### 2. Preparation of the residue for combustion—

The combustion tubing, which should be of rather smaller bore than that usually employed in organic analysis, is broken into pieces about 18 inches in length, and after thoroughly cleansing and drying one end is sealed up. A layer, about 1 inch in length, of the coarse "assayed" oxide of copper (for preparation see below) is introduced into the tube. To the residue in the dish a little fine oxide of copper is added, and the two are intimately mixed by means of a small flexible steel spatula, by means of which the residue can be readily detached from the dish. By means of a small scoop of copper-foil this mixture is then tilted into the tube, and the dish is similarly rinsed out once or twice more with fresh quantities of fine oxide of copper. The tube is now filled to a distance of 10 inches from the closed end with coarse oxide, which is well shaken down by tapping the closed end of the tube on the table. Above the oxide of copper is now placed a cylinder of copper-gauze 3 inches long, and beyond this again an inch of oxide of copper.

whole of the ammonia is in this way liberated, but a separate determination of the organic carbon must be made by evaporating another portion of the water with sulphurous acid, inasmuch as carbonates are not wholly decomposed by boracic acid.

When charged as above, the open end of the tube is drawn out before the blowpipe to a stout tube about 6 inches long and ½ in diameter. The end of this tube is rounded off, and the long neck bent at right angles in the middle.

### 3. The combustion—

The tube, prepared as above, is placed in an ordinary combustion-furnace, and the open end connected with a Sprengel pump, as shown in Fig. 12. The arrangement of the pump is sufficiently evident from the figure, and requires no further explanation excepting that the junction between the two tubes a and c is effected by means of caoutchouc, surrounded by a small piece of wide glass tubing filled with glycerine to prevent the possibility of any leakage of air. The joint which connects the open end of the combustion-tube with the pump is of a precisely similar character, only that water is substituted for glycerine.

When the connection has been made, the pump is set in action by carefully opening the clip (d), and the exhaustion of the combustion-tube commences. Whilst this is going on the front burners of the furnace are lit so as to heat the copper-cylinder and about one inch of oxide, the further heating of the tube being prevented by placing an iron screen across the furnace. The tin trough (b) is during this time filled with hot water, so as to evaporate any water in the bulb that may be left from a previous operation.

In about ten minutes' time the exhaustion of the tube will be indicated by the metallic click with which the falling pistons of mercury strike upon the column of mercury, now about 30 inches in height, in the tube (B).

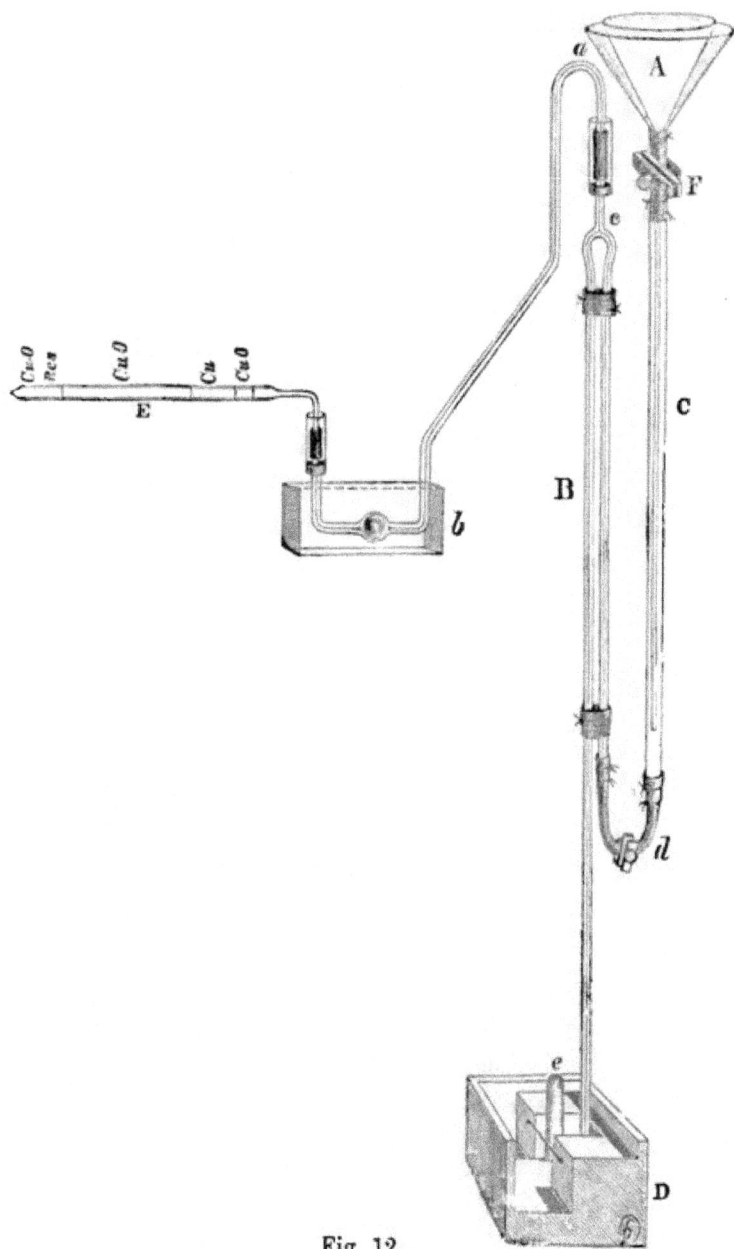

Fig. 12.

The combustion of the residue is now proceeded with by turning on the remaining burners gradually until the whole length of the tube is at a dull red-heat. During this time the hot water in the trough (*b*) is exchanged for cold, in order that the water produced during the combustion may be condensed there instead of passing on into the small test-tube (*e*), which is filled with mercury and placed over the lower extremity of the descending tube (*B*) for the reception of the gases.

The combustion is known to be complete when the column of mercury in the tube (*B*) ceases to fall. The combustion is usually finished in one hour.

The pump is now again set in action by opening the clamp (*d*), so as to transfer the gaseous products of combustion into the test-tube (*e*). The heat is not reduced during the exhaustion lest the tube should crack and admit air.

When the exhaustion is complete, the tube (*e*) is removed and its contents subjected to gasometric measurement.

### 4. The volumetric measurement of the gases—

The measurement of the gases is conducted in the apparatus shown in Fig. 13.

"It consists of a U-shaped tube with unequal limbs. In the shorter limb (*A*) the volume of the gas under examination is ascertained, and in the longer limb (*B*) the pressure to which the gas is subjected is measured. By turning a glass stopcock (*a*) the closed limb (*A*) can be put in connection with a chamber (*C*) called the "laboratory-tube," in which the gas undergoes chemical treatment. By lowering or raising the vessel (*D*) of mercury connected by a flexible tube (*c*) with the foot of the longer limb, mercury is withdrawn or forced into the measuring apparatus; and

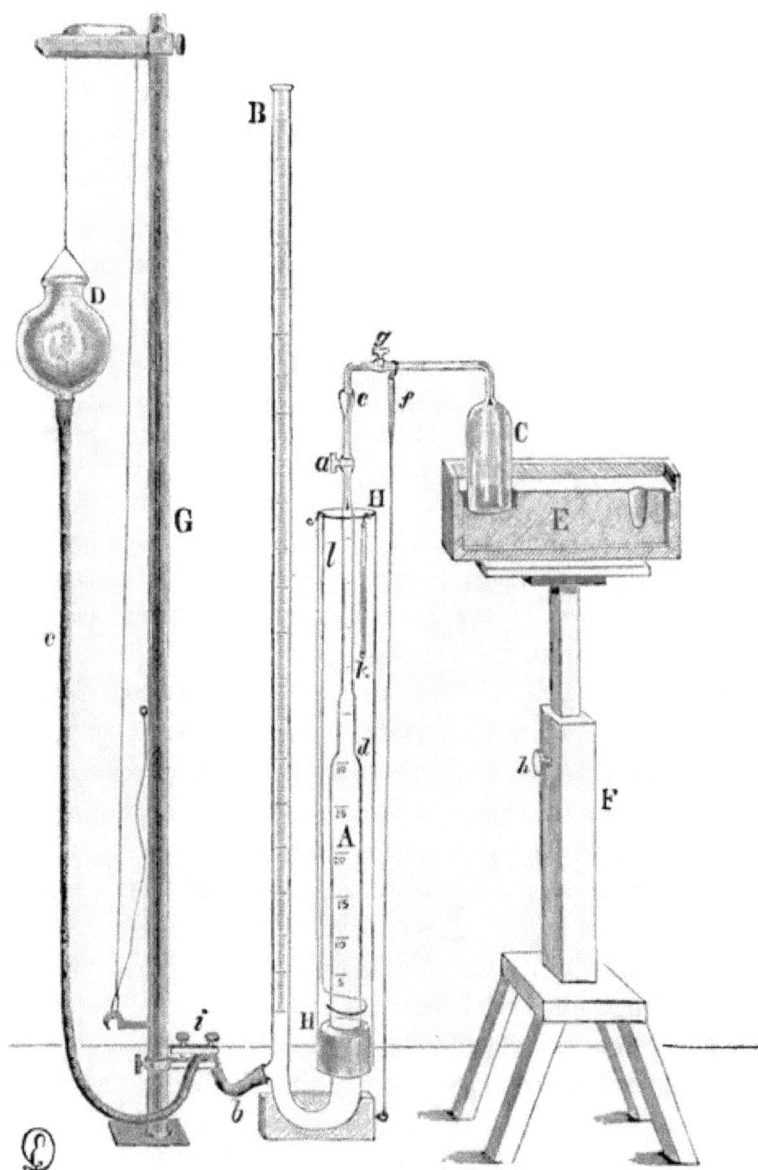

Fig. 13.

by this means the motive power is provided which draws
gas from the "laboratory-tube" into the measuring tube,
expels it again, and regulates the pressure during measurement.    The measuring chamber or shorter limb is a stout
glass-tube of ·75 inch internal diameter, bent at its lower
end and joined to the pressure-tube (*B*), 4 feet long and of
·25 inch internal diameter.    Near the bottom of (*B*) is
inserted a glass-tube (*b*), 1 inch long and ·2 inch in diameter, for the admission and withdrawal of mercury.    It
is connected with the movable mercury reservoir (*D*) by
the webbed caoutchouc-tube (*c*).

"In order to permit of both small and large quantities
of gas being measured with equal precision, the tube (*A*)
decreases in bore towards the top.    The measuring chamber
terminates above in a capillary tube provided with a glass
stopcock (*a*)."

The measuring tube is connected with the laboratory
vessel by means of the joint shown at *e*, which has the
following arrangement.    About an inch above the stopcock
on *A* the capillary tube is expanded into a small funnel
or cup (*e*).    The capillary tube from the laboratory vessel
is bent twice at right angles, and then drawn out at its
end so as to fit into the neck of the cup.    The joint is
rendered gas-tight by fitting a piece of thin india-rubber
tubing into the narrower part of the cup, and then keeping
the conical stopper of the laboratory-vessel-tube firmly
fixed in this by means of the elastic band (*f*); the cup
itself is then filled with mercury.

The laboratory vessel (*C*), which is 4 inches in height
and 1½ inch internal diameter, stands firmly on the shelf
of a strong mahogany mercury-trough (*E*), the construction
of which is shown in the figure, and in Fig. 14.

The measuring tube is surrounded by a larger glass

tube (*II II*), containing water to give a definite temperature
to the gases during measurement, and a thermometer (*k*)
hangs in this water-jacket. The uniform temperature of
the water is secured by means of an agitator (*l*) before each
reading. The two limbs of the U-shaped-tube are so
graduated that the zeros on both are at the same level.
The pressure tube is graduated into single millimetres, the
measuring tube at intervals of 10 m.m., commencing at
about 2 inches above the bend or foot.

Before the apparatus can be used, the measuring tube

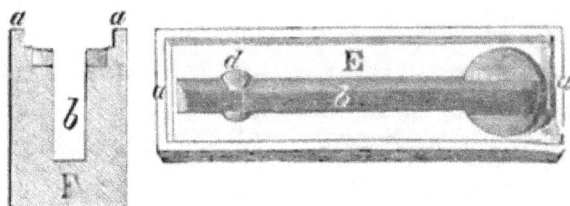

Fig. 14.

must once and for all be **calibrated**, and the **corrections
for capillarity** in the two tubes determined.

The calibration, or determination of the capacity of the
measuring tube as the mercury stands at each graduation,
is effected by fitting a small piece of capillary tubing, bent
twice at right angles, to the cup of the measuring tube, in
the same manner as the tube of the laboratory vessel is
fitted. By raising the reservoir (*D*) the mercury is made
to drop from the open extremity of this capillary tube; the
latter is then immersed in a beaker of distilled water, and
the measuring tube filled with water by lowering the
reservoir. When the water in the measuring tube is below
the zero, the beaker is removed and the mercury reservoir
raised until the surface of the mercury stands exactly at

the zero, which is best determined by means of a small telescope sliding on a vertical rod. The temperature of the water in the jacket is now noted, and the calibration commenced. A small tared flask is so placed as to receive the water that falls from the capillary tube on making the mercury rise in the measuring tube. By raising the reservoir (D) the surface of the mercury is made to rise from zero to the first graduation exactly, and the flask containing the water that has escaped during this operation is weighed again. The same operation is repeated for each graduation on the tube. If the temperature of the water, as indicated by that in the jacket, is 4° C., the weight in grms. of the water expelled during each of these operations gives the capacity in c.c. corresponding to the respective lengths of the tube. From this a table is constructed showing the volume of the measuring tube and the logarithmic equivalent of this volume, as the mercury stands at each division in the tube.

The corrections for capillarity are made by taking readings in the two limbs, for each graduation in the measuring tube, when both are freely open to the air. The positive or negative correction which the readings of pressures will require must likewise be inserted in the above table.

## ANALYSIS OF GASES.

Before using the apparatus the stopcocks of both measuring tube and laboratory vessel must be lightly greased with cerate, and the measuring tube filled with mercury by raising the reservoir until the cup begins to fill. The laboratory vessel is also filled with mercury by exhausting the air by suction. The laboratory vessel and measuring tube are now placed in connection as already described, and then the apparatus is ready for use.

The gaseous products of combustion contained in the test-tube, in which they were collected by means of the Sprengel pump, are transferred to the laboratory vessel. A few drops of a solution of potassic dichromate are added by means of a bent pipette to absorb any sulphurous acid, which is rarely present in the gases; should the chromate turn green, however, more must be added. The thorough contact of the gases with the liquid reagent is best ensured by allowing a gentle stream of mercury to flow into the laboratory vessel by keeping the reservoir ($D$) above the level of the junction. The absorption, if any, will be complete in two minutes.

The gas is now drawn over into the measuring tube by lowering the reservoir ($D$), the stopcock ($g$) is closed just as the liquid reageant approaches it, and the cock ($a$) immediately afterwards. The level of the mercury in the measuring tube is then made to coincide with one of the graduations, and the height of the mercury in the two tubes read off by means of the telescope, together with the height of the barometer at the time, and the temperature of the water in the jacket. From these observations the total volume ($A$) of the gas is calculated.

The gas is now returned to the laboratory vessel, into which a few drops of a saturated solution of caustic potash have been introduced by means of a pipette, in order to remove the carbonic acid, which is absorbed within a space of five minutes. The residual gas is then again measured, a second reading of the barometer being, however, unnecessary. Let the volume obtained be denoted by $B$.

The gas is now again returned into the laboratory vessel,

into which a few drops (not more than one-half volume of the potash) of a saturated solution of pyrogallic acid have been introduced.   If the liquid runs off the glass without a stain, showing the absence of free oxygen, a bubble or two of pure oxygen is admitted, and the mercury caused to flow slowly into the laboratory vessel for about ten minutes. The object of the addition of oxygen is to cause the absorption of a small quantity of nitric oxide, which is generally present in the gas, and which has escaped the reducing action of the copper-cylinder.   By measuring the gas again, the diminution in volume gives the nitric oxide, and the remaining gas is nitrogen, which may be denoted by $C$.   From the above analysis we have :—

$A$ = volume of carbonic anhydride, nitric oxide, and nitrogen.

$B$ = volume of nitric oxide and nitrogen.

$C$ = volume of nitrogen.

Let these volumes calculated to 0° C. and 760 m.m. press., be denoted by A′, B′, and C′, respectively.   The volumes of carbonic anhydride and nitrogen may then be obtained by the equations :—

$$A' - B' = \text{vol. of } CO_2$$
$$\frac{B' - C'}{2} + C' = \frac{B' + C'}{2} = \text{vol. of N,}$$

and from these the weights of carbon and nitrogen can be readily calculated.   This result may, however, be much more rapidly arrived at by dispensing with the inter- mediate determination of the corrected volumes, and assuming that the original gaseous mixture consists entirely of nitrogen.   Then taking into consideration (1) that the weights of carbon and nitrogen contained in equal volumes of carbonic anhydride and nitrogen measured at the same pressure and temperature, are to each other as 6 : 14, and (2) that the weights of nitrogen contained in equal volumes of nitrogen and nitric oxide are as 2 : 1, then, if A be the weight of the total gas calculated as nitrogen, B the weight after absorption of the first portion $(CO_2)$, and C the weight after the absorption of the second portion $(N_2O_2)$; further, if $x$ and $y$ represent respectively the weights of carbon and nitrogen actually contained in the gaseous mixture, then the following equations express the values of $x$ and $y$ :—

$$x = \frac{3(A - B)}{7}$$

$$y = \frac{C + B}{2}$$

By the use of the Logarithmic Table given below for the reduction of cubic centimetres of nitrogen to grms., for each tenth of a degree centigrade, by the formula $\frac{·0012562}{(1 + ·00367t)760}$, the labour of calculation is reduced to a minimum. The following example will make the method of calculation more intelligible :—

### Carbon and Nitrogen Determination in 1 Litre of Water.

| A. Total Volume of Gas after contact with Bichromate of Potash. | | Division on Measuring Tube. | Tempera-ture. | Barometer. |
|---|---|---|---|---|
| Height of mercury | | | | |
| in measuring tube | 208·9 m.m. | 30 = 3·334 c.c. | 20·4 C. | 753·0 m.m. |
| „ „ pressure tube | 130·0 „ | ... | ... | 95·8 „ |
| Difference . . . | 78·9 „ | Pressure on dry gas at | | |
| Plus tension of | | 20·4 C. . . . . . | | 657·2 „ |
| aqueous vapour . | 17·8 „ | | | |
| Minus correction for | | | | |
| capillarity. . . | − ·9 „ | | | |
| | 95·8 „ | to be deducted from height of barometer. | | |

| B. Volume after absorption of $CO_2$ by KOH. | Division on Measuring Tube. | Temperature. | Barometer. |
|---|---|---|---|
| Measuring tube . . 426·1 m.m. | 50 = ·361 c.c. | 20°·4 C. | 753·0 m.m. |
| Pressure tube . . 229·9 „ | ... | ... | 216·2 „ |
| Difference . . . 196·2 „ | Pressure on dry gas at | | |
| Plus tension of | 20°·4 C. . . . . 536·8 „ | | |
| aqueous vapour . 17·8 „ | | | |
| Plus correction for capillarity, division 50 being on narrowest part of tube . . . . 2·2 „ | | | |

216·2 „ to be deducted from height of barometer.

| C. Volume after absorption of $N_2O_2$ by Oxygen and Potassic Pyrogallate. | Division on Measuring-Tube. | Temperature. | Barometer. |
|---|---|---|---|
| Measuring tube . . 426·1 m.m. | 50 = ·361 c.c. | 20°·3 C. | 753·0 m.m. |
| Pressure tube . . 190·4 „ | ... | ... | 255·6 „ |
| Difference . . . 235·7 „ | Pressure on dry gas at | | |
| Plus tension of | 20°·3 C. . . . . 497·4 „ | | |
| aqueous vapour . 17·7 „ | | | |
| Plus correction for capillarity . . . 2·2 „ | | | |

255·6 „ to be deducted from height of barometer.

| | A. | B. | C. |
|---|---|---|---|
| Log. of volume corresponding to division on measuring tube . | 0·52290 | $\bar{1}$·55766 | $\bar{1}$·55766 |
| Log. of $\dfrac{0·0012562}{(1+0·00367t)760}$ (see table below) | $\bar{6}$·18688 | $\bar{6}$·18688 | $\bar{6}$·18703 |
| Log. of pressure on dry gas . . | 2·81769 | 2·72981 | 2·69670 |
| Log. of weight of gas calculated as nitrogen . . . . . . . | $\bar{3}$·52747 | $\bar{4}$·47435 | $\bar{4}$·44139 |
| Natural number . . . . = | ·0033688 | ·0002980 | ·0002763 |

Now, by applying the formulæ on p. 307 :—

$$\cdot 0033688 = A.$$
$$\cdot 0002980 = B.$$

---

$$\cdot 0030708$$
$$3$$

---

$$7)\cdot 0092124$$

---

$$\cdot 0013160 = \text{weight of carbon.}$$

$$= \cdot 132 \text{ part per 100,000.}$$
$$- \cdot 006 \text{ correction for blank combustion.}$$

---

Organic carbon $= \cdot 126$ part per 100,000.

$$\cdot 0002980 = B.$$
$$\cdot 0002763 = C.$$

---

$$2)\cdot 0005743$$

---

$$\cdot 0002871 = \text{weight of nitrogen.}$$

$$= \cdot 029 \text{ part per 100,000.}$$
$$- \cdot 005 \text{ correction for blank combustion.}$$

---

Nitrogen $= \cdot 024$ part per 100,000.
$$- \cdot 003 \text{ deduction for N as } NH_3$$

---

Organic nitrogen $= \cdot 021$ part per 100,000.

As already stated, and as seen in the above example, the nitrogen corresponding to the ammonia found in the water must be subtracted from the nitrogen obtained by combustion. The ammonic sulphite, however, during the evaporation of the water, incurs a slight loss of ammonia ; and hence in waters containing more than $\cdot 01$ part of ammonia per 100,000, this loss must be taken into account, the nitrogen to be subtracted being found in the table given below. The salt of ammonia, which, during the evaporation of its solution, suffers least loss of ammonia, is ammonic phosphate; and hence when a water contains a large proportion of ammonia and no nitrates, it is preferable to substitute a solution of metaphosphoric acid for one of sulphurous acid in the evaporation of the water.

## Loss of Nitrogen by Evaporation of $NH_3(SO(HO)_2)$.

### Parts in 100,000.

| $NH_3$ | Loss of N. | $NH_3$ | Loss of N. | $NH_3$ | Loss of N. | $NH_3$ | Loss of N. | $NH_3$ | Loss of N. | $NH_3$ | Loss of N. |
|---|---|---|---|---|---|---|---|---|---|---|---|
| 6·0 | 1·727 | 4·8 | 1·451 | 3·6 | ·977 | 2·4 | ·503 | 1·2 | ·250 | ·09 | ·014 |
| 5·9 | 1·707 | 4·7 | 1·411 | 3·5 | ·937 | 2·3 | ·463 | 1·1 | ·238 | ·08 | ·013 |
| 5·8 | 1·688 | 4·6 | 1·372 | 3·4 | ·898 | 2·2 | ·424 | 1·0 | ·226 | ·07 | ·012 |
| 5·7 | 1·668 | 4·5 | 1·332 | 3·3 | ·858 | 2·1 | ·384 | ·9 | ·196 | ·06 | ·010 |
| 5·6 | 1·648 | 4·4 | 1·293 | 3·2 | ·819 | 2·0 | ·345 | ·8 | ·166 | ·05 | ·009 |
| 5·5 | 1·628 | 4·3 | 1·253 | 3·1 | ·779 | 1·9 | ·333 | ·7 | ·136 | ·04 | ·007 |
| 5·4 | 1·609 | 4·2 | 1·214 | 3·0 | ·740 | 1·8 | ·321 | ·6 | ·106 | ·03 | ·006 |
| 5·3 | 1·589 | 4·1 | 1·174 | 2·9 | ·700 | 1·7 | ·309 | ·5 | ·077 | ·02 | ·004 |
| 5·2 | 1·569 | 4·0 | 1·135 | 2·8 | ·661 | 1·6 | ·297 | ·4 | ·062 | ·01 | ·003 |
| 5·1 | 1·549 | 3·9 | 1·095 | 2·7 | ·621 | 1·5 | ·285 | ·3 | ·047 | ·009 | ·001 |
| 5·0 | 1·530 | 3·8 | 1·056 | 2·6 | ·582 | 1·4 | ·274 | ·2 | ·032 | ... | ... |
| 4·9 | 1·490 | 3·7 | 1·016 | 2·5 | ·542 | 1·3 | ·262 | ·1 | ·017 | ... | .. |

## LOSS OF NITROGEN BY EVAPORATION OF $NH_3(PO_2(HO))$:—

| Volume evaporated. | NH₃ per 100,000. | Loss of N per 100,000. | Volume evaporated. | NH₃ per 100,000. | Loss of N per 100,000. | Volume evaporated. | NH₃ per 100,000. | Loss of N per 100,000. | Volume evaporated. | NH₃ per 100,000. | Loss of N per 100,000. |
|---|---|---|---|---|---|---|---|---|---|---|---|
| c.c. | | | c.c. | | | c.c. | | | c.c. | | |
| 100 | 10·0 | ·483 | 100 | 7·2 | ·386 | 100 | 4·5 | ·287 | 100 | 1·8 | ·153 |
| ,, | 9·9 | ·480 | ,, | 7·1 | ·382 | ,, | 4·4 | ·283 | ,, | 1·7 | ·148 |
| ,, | 9·8 | ·476 | ,, | 7·0 | ·379 | ,, | 4·3 | ·279 | ,, | 1·6 | ·143 |
| ,, | 9·7 | ·473 | ,, | 6·9 | ·375 | ,, | 4·2 | ·275 | ,, | 1·5 | ·137 |
| ,, | 9·6 | ·469 | ,, | 6·8 | ·372 | ,, | 4·1 | ·271 | ,, | 1·4 | ·132 |
| ,, | 9·5 | ·466 | ,, | 6·7 | ·368 | ,, | 4·0 | ·267 | ,, | 1·3 | ·127 |
| ,, | 9·4 | ·462 | ,, | 6·6 | ·365 | ,, | 3·9 | ·262 | ,, | 1·2 | ·122 |
| ,, | 9·3 | ·459 | ,, | 6·5 | ·361 | ,, | 3·8 | ·257 | ,, | 1·1 | ·117 |
| ,, | 9·2 | ·455 | ,, | 6·4 | ·358 | ,, | 3·7 | ·252 | ,, | 1·0 | ·112 |
| ,, | 9·1 | ·452 | ,, | 6·3 | ·354 | ,, | 3·6 | ·247 | 250 | 0·9 | ·096 |
| ,, | 9·0 | ·448 | ,, | 6·2 | ·351 | ,, | 3·5 | ·242 | ,, | 0·8 | ·080 |
| ,, | 8·9 | ·445 | ,, | 6·1 | ·348 | ,, | 3·4 | ·236 | ,, | 0·7 | ·070 |
| ,, | 8·8 | ·441 | ,, | 6·0 | ·345 | ,, | 3·3 | ·231 | ,, | 0·6 | ·060 |
| ,, | 8·7 | ·438 | ,, | 5·9 | ·341 | ,, | 3·2 | ·226 | 500 | 0·5 | ·050 |
| ,, | 8·6 | ·434 | ,, | 5·8 | ·337 | ,, | 3·1 | ·221 | ,, | 0·4 | ·040 |
| ,, | 8·5 | ·431 | ,, | 5·7 | ·333 | ,, | 3·0 | ·216 | ,, | 0·3 | ·030 |
| ,, | 8·4 | ·428 | ,, | 5·6 | ·330 | ,, | 2·9 | ·211 | 1000 | 0·2 | ·020 |
| ,, | 8·3 | ·424 | ,, | 5·5 | ·326 | ,, | 2·8 | ·205 | ,, | 0·1 | ·010 |
| ,, | 8·2 | ·421 | ,, | 5·4 | ·322 | ,, | 2·7 | ·200 | ,, | 0·09 | ·009 |
| ,, | 8·1 | ·417 | ,, | 5·3 | ·318 | ,, | 2·6 | ·195 | ,, | 0·08 | ·008 |
| ,, | 8·0 | ·414 | ,, | 5·2 | ·314 | ,, | 2·5 | ·190 | ,, | 0·07 | ·007 |
| ,, | 7·9 | ·410 | ,, | 5·1 | ·310 | ,, | 2·4 | ·184 | ,, | 0·06 | ·006 |
| ,, | 7·8 | ·407 | ,, | 5·0 | ·306 | ,, | 2·3 | ·179 | ,, | 0·05 | ·005 |
| ,, | 7·7 | ·403 | ,, | 4·9 | ·302 | ,, | 2·2 | ·174 | ,, | 0·04 | ·004 |
| ,, | 7·6 | ·400 | ,, | 4·8 | ·298 | ,, | 2·1 | ·169 | ,, | 0·03 | ·003 |
| ,, | 7·5 | ·396 | ,, | 4·7 | ·294 | ,, | 2·0 | ·164 | ,, | 0·02 | ·002 |
| ,, | 7·4 | ·393 | ,, | 4·6 | ·291 | ,, | 1·9 | ·158 | ,, | 0·01 | ·001 |
| ,, | 7·3 | ·389 | ... | ... | ... | ... | ... | ... | ... | ... | ... |

## REDUCTION OF CUBIC CENTIMETRES OF NITROGEN TO GRAMMES.

Log. $\dfrac{0 \cdot 0012562}{(1 + 0 \cdot 00367t)760}$   for each tenth of a degree from 0° to 30° C.

| T.C. | 0·0 | 0·1 | 0·2 | 0·3 | 0·4 | 0·5 | 0·6 | 0·7 | 0·8 | 0·9 |
|---|---|---|---|---|---|---|---|---|---|---|
| 0° | 6·21824 | 808 | 793 | 777 | 761 | 745 | 729 | 713 | 697 | 681 |
| 1 | 665 | 649 | 633 | 617 | 601 | 586 | 570 | 554 | 538 | 522 |
| 2 | 507 | 491 | 475 | 459 | 443 | 427 | 412 | 396 | 380 | 364 |
| 3 | 349 | 333 | 318 | 302 | 286 | 270 | 255 | 239 | 223 | 208 |
| 4 | 192 | 177 | 161 | 145 | 130 | 114 | 098 | 083 | 067 | 051 |
| 5 | 035 | 020 | 004 | *989 | *973 | *957 | *942 | *926 | *911 | *895 |
| 6 | 6·20879 | 864 | 848 | 833 | 817 | 801 | 786 | 770 | 755 | 739 |
| 7 | 723 | 708 | 692 | 676 | 661 | 645 | 629 | 614 | 598 | 583 |
| 8 | 567 | 552 | 536 | 521 | 505 | 490 | 474 | 459 | 443 | 428 |
| 9 | 413 | 397 | 382 | 366 | 351 | 335 | 320 | 304 | 289 | 274 |
| 10 | 259 | 244 | 228 | 213 | 198 | 182 | 167 | 151 | 136 | 121 |
| 11 | 106 | 090 | 075 | 060 | 045 | 029 | 014 | *999 | *984 | *969 |
| 12 | 6·19953 | 938 | 923 | 907 | 892 | 877 | 862 | 846 | 831 | 816 |
| 13 | 800 | 785 | 770 | 755 | 740 | 724 | 709 | 694 | 679 | 664 |
| 14 | 648 | 633 | 618 | 603 | 588 | 573 | 558 | 543 | 528 | 513 |
| 15 | 497 | 482 | 467 | 452 | 437 | 422 | 407 | 392 | 377 | 362 |
| 16 | 346 | 331 | 316 | 301 | 286 | 271 | 256 | 241 | 226 | 211 |
| 17 | 196 | 181 | 166 | 151 | 136 | 121 | 106 | 091 | 076 | 061 |
| 18 | 046 | 031 | 016 | 001 | *986 | *971 | *956 | *941 | *926 | *911 |
| 19 | 6·18897 | 882 | 867 | 852 | 837 | 822 | 807 | 792 | 777 | 762 |
| 20 | 748 | 733 | 718 | 703 | 688 | 673 | 659 | 644 | 629 | 614 |
| 21 | 600 | 585 | 570 | 555 | 540 | 526 | 511 | 496 | 481 | 466 |
| 22 | 452 | 437 | 422 | 408 | 393 | 378 | 363 | 349 | 334 | 319 |
| 23 | 305 | 290 | 275 | 261 | 246 | 231 | 216 | 202 | 187 | 172 |
| 24 | 158 | 143 | 128 | 114 | 099 | 084 | 070 | 055 | 041 | 026 |
| 25 | 012 | *997 | *982 | *968 | *953 | *938 | *924 | *909 | *895 | *880 |
| 26 | 6·17866 | 851 | 837 | 822 | 808 | 793 | 779 | 764 | 750 | 735 |
| 27 | 721 | 706 | 692 | 677 | 663 | 648 | 634 | 619 | 605 | 590 |
| 28 | 576 | 561 | 547 | 532 | 518 | 503 | 489 | 475 | 460 | 446 |
| 29 | 432 | 417 | 403 | 388 | 374 | 360 | 345 | 331 | 316 | 302 |

# Tension of Aqueous Vapour for each $\frac{1}{10}$th Degree Centigrade, from 0° to 30° C. (Regnault.)

| Temp. C. | Tension in m.m. of mercury. | Temp. C. | Tension in m.m. of mercury. | Temp. C. | Tension in m.m. of mercury. | Temp. C. | Tension in m.m. of mercury. | Temp. C. | Tension in m.m. of mercury. |
|---|---|---|---|---|---|---|---|---|---|
| 0   | 4·6 | 6·0  | 7·0  | 12·0 | 10·5 | 18·0 | 15·4 | 24·0 | 22·2 |
| ·1  | 4·6 | ·1   | 7·0  | ·1   | 10·5 | ·1   | 15·5 | ·1   | 22·3 |
| ·2  | 4·7 | ·2   | 7·1  | ·2   | 10·6 | ·2   | 15·6 | ·2   | 22·5 |
| ·3  | 4·7 | ·3   | 7·1  | ·3   | 10·7 | ·3   | 15·7 | ·3   | 22·6 |
| ·4  | 4·7 | ·4   | 7·2  | ·4   | 10·7 | ·4   | 15·7 | ·4   | 22·7 |
| ·5  | 4·8 | ·5   | 7·2  | ·5   | 10·8 | ·5   | 15·8 | ·5   | 22·9 |
| ·6  | 4·8 | ·6   | 7·3  | ·6   | 10·9 | ·6   | 15·9 | ·6   | 23·0 |
| ·7  | 4·8 | ·7   | 7·3  | ·7   | 10·9 | ·7   | 16·0 | ·7   | 23·1 |
| ·8  | 4·9 | ·8   | 7·4  | ·8   | 11·0 | 8    | 16·1 | ·8   | 23·3 |
| ·9  | 4·9 | ·9   | 7·4  | ·9   | 11·1 | ·9   | 16·2 | ·9   | 23·4 |
| 1·0 | 4·9 | 7·0  | 7·5  | 13·0 | 11·2 | 19·0 | 16·3 | 25·0 | 23·5 |
| ·1  | 5·0 | ·1   | 7·5  | ·1   | 11·2 | ·1   | 16·4 | ·1   | 23·7 |
| ·2  | 5·0 | ·2   | 7·6  | ·2   | 11·3 | ·2   | 16·6 | ·2   | 23·8 |
| ·3  | 5·0 | ·3   | 7·6  | ·3   | 11·4 | ·3   | 16·7 | ·3   | 24·0 |
| ·4  | 5·1 | ·4   | 7·7  | ·4   | 11·5 | ·4   | 16·8 | ·4   | 24·1 |
| ·5  | 5·1 | ·5   | 7·8  | ·5   | 11·5 | ·5   | 16·9 | ·5   | 24·3 |
| ·6  | 5·2 | ·6   | 7·8  | ·6   | 11·6 | ·6   | 17·0 | ·6   | 24·4 |
| ·7  | 5·2 | ·7   | 7·9  | ·7   | 11·7 | ·7   | 17·1 | ·7   | 24·6 |
| ·8  | 5·2 | ·8   | 7·9  | ·8   | 11·8 | ·8   | 17·2 | ·8   | 24·7 |
| ·9  | 5·3 | ·9   | 8·0  | ·9   | 11·8 | ·9   | 17·3 | ·9   | 24·8 |
| 2·0 | 5·3 | 8·0  | 8·0  | 14·0 | 11·9 | 20·0 | 17·4 | 26·0 | 25·0 |
| ·1  | 5·3 | ·1   | 8·1  | ·1   | 12·0 | ·1   | 17·5 | ·1   | 25·1 |
| ·2  | 5·4 | ·2   | 8·1  | ·2   | 12·1 | ·2   | 17·6 | ·2   | 25·3 |
| ·3  | 5·4 | ·3   | 8·2  | ·3   | 12·1 | ·3   | 17·7 | ·3   | 25·4 |
| ·4  | 5·5 | ·4   | 8·2  | ·4   | 12·2 | ·4   | 17·8 | ·4   | 25·6 |
| ·5  | 5·5 | ·5   | 8·3  | ·5   | 12·3 | ·5   | 17·9 | ·5   | 25·7 |
| ·6  | 5·5 | ·6   | 8·3  | ·6   | 12·4 | ·6   | 18·0 | ·6   | 25·9 |
| ·7  | 5·6 | ·7   | 8·4  | ·7   | 12·5 | ·7   | 18·2 | ·7   | 26·0 |
| ·8  | 5·6 | ·8   | 8·5  | ·8   | 12·5 | ·8   | 18·3 | ·8   | 26·2 |
| ·9  | 5·6 | ·9   | 8·5  | ·9   | 12·6 | ·9   | 18·4 | ·9   | 26·4 |
| 3·0 | 5·7 | 9·0  | 8·6  | 15·0 | 12·7 | 21·0 | 18·5 | 27·0 | 26·5 |
| ·1  | 5·7 | ·1   | 8·6  | ·1   | 12·8 | ·1   | 18·6 | ·1   | 26·7 |
| ·2  | 5·8 | ·2   | 8·7  | ·2   | 12·9 | ·2   | 18·7 | ·2   | 26·8 |
| ·3  | 5·8 | ·3   | 8·7  | ·3   | 12·9 | ·3   | 18·8 | ·3   | 27·0 |
| ·4  | 5·8 | ·4   | 8·8  | ·4   | 13·0 | ·4   | 19·0 | ·4   | 27·1 |
| ·5  | 5·9 | ·5   | 8·9  | ·5   | 13·1 | ·5   | 19·1 | ·5   | 27·3 |
| ·6  | 5·9 | ·6   | 8·9  | ·6   | 13·2 | ·6   | 19·2 | ·6   | 27·5 |
| ·7  | 6·0 | ·7   | 9·0  | ·7   | 13·3 | ·7   | 19·3 | ·7   | 27·6 |
| ·8  | 6·0 | ·8   | 9·0  | ·8   | 13·4 | ·8   | 19·4 | ·8   | 27·8 |
| ·9  | 6·1 | ·9   | 9·1  | ·9   | 13·5 | ·9   | 19·5 | ·9   | 27·9 |
| 4·0 | 6·1 | 10·0 | 9·2  | 16·0 | 13·5 | 22·0 | 19·7 | 28·0 | 28·1 |
| ·1  | 6·1 | ·1   | 9·2  | ·1   | 13·6 | ·1   | 19·8 | ·1   | 28·3 |
| ·2  | 6·2 | ·2   | 9·3  | ·2   | 13·7 | ·2   | 19·9 | ·2   | 28·4 |
| ·3  | 6·2 | ·3   | 9·3  | ·3   | 13·8 | ·3   | 20·0 | ·3   | 28·6 |
| ·4  | 6·3 | ·4   | 9·4  | ·4   | 13·9 | ·4   | 20·1 | ·4   | 28·8 |
| ·5  | 6·3 | ·5   | 9·5  | ·5   | 14·0 | ·5   | 20·3 | ·5   | 28·9 |
| ·6  | 6·4 | ·6   | 9·5  | ·6   | 14·1 | ·6   | 20·4 | ·6   | 29·1 |
| ·7  | 6·4 | ·7   | 9·6  | ·7   | 14·2 | ·7   | 20·5 | ·7   | 29·3 |
| ·8  | 6·4 | ·8   | 9·7  | ·8   | 14·2 | ·8   | 20·6 | ·8   | 29·4 |
| ·9  | 6·5 | ·9   | 9·7  | ·9   | 14·3 | ·9   | 20·8 | ·9   | 29·6 |
| 5·0 | 6·5 | 11·0 | 9·8  | 17·0 | 14·4 | 23·0 | 20·9 | 29·0 | 29·8 |
| ·1  | 6·6 | ·1   | 9·9  | ·1   | 14·5 | ·1   | 21·0 | ·1   | 30·0 |
| ·2  | 6·6 | ·2   | 9·9  | ·2   | 14·6 | ·2   | 21·1 | ·2   | 30·1 |
| ·3  | 6·7 | ·3   | 10·0 | ·3   | 14·7 | ·3   | 21·3 | ·3   | 30·3 |
| ·4  | 6·7 | ·4   | 10·1 | ·4   | 14·8 | ·4   | 21·4 | ·4   | 30·5 |
| ·5  | 6·8 | ·5   | 10·1 | ·5   | 14·9 | ·5   | 21·5 | ·5   | 30·7 |
| ·6  | 6·8 | ·6   | 10·2 | ·6   | 15·0 | ·6   | 21·7 | ·6   | 30·8 |
| ·7  | 6·9 | ·7   | 10·3 | ·7   | 15·1 | ·7   | 21·8 | ·7   | 31·0 |
| ·8  | 6·9 | ·8   | 10·3 | ·8   | 15·2 | ·8   | 21·9 | ·8   | 31·2 |
| ·9  | 7·0 | ·9   | 10·4 | ·9   | 15·3 | ·9   | 22·1 | ·9   | 31·4 |

**Preparation of solution of sulphurous acid.**—The sulphurous anhydride must be prepared from pure re-distilled sulphuric acid and clean copper cuttings. The copper should be previously digested for twenty-four hours with concentrated sulphuric acid, and then washed with water before use. The sulphurous anhydride is passed through a washbottle containing a little water, and then into water free from ammonia and organic matter, until the solution is saturated.

**Preparation of solution of hydric sodic sulphite.**— Ignited sodic carbonate is dissolved in water free from organic matter and ammonia, and then sulphurous anhydride, prepared as above, passed into the solution until carbonic anhydride ceases to be evolved.

**Preparation of oxygen for gas analysis.**—A bulb of about 20 c.c. capacity is blown on the end of a piece of combustion tubing; this is filled with dried and coarsely-powdered potassic chlorate, and the tube is then drawn out and bent at an angle, so that the whole forms a little retort with long delivery-tube. The end of the delivery-tube is made to dip below the surface of mercury, and the bulb is then heated. The first 80 c.c. of the gas evolved are rejected, the remainder is then collected in clean test-tubes over mercury. The gas should leave but a very minute bubble after treatment with potassic hydrate and pyrogallic acid.

**Preparation of solution of metaphosphoric acid.**— 100 grms. of the glacial acid are dissolved in 1 litre of organically pure water; 10 c.c. should not contain an appreciable amount of ammonia.

**Preparation of calcic phosphate.**—To a solution of calcic chloride, a solution of hydric disodic phosphate is added. The precipitate is washed with water by decantation, dried, and afterwards heated to redness for one hour.

**Oxide of copper.**—This should not have been treated with nitric acid, but have been calcined for two or three hours either in a muffle or in an iron-tube, open at both ends, placed in a combustion-furnace tilted so as to cause a current of air to pass through the tube. A blank combustion must be performed with the product of each calcination; only a minute bubble of gas should be obtained, and this should be almost wholly absorbed by caustic potash. The carbonic anhydride found should not correspond to more than ·00005 grm. of carbon.

**Copper-gauze cylinders.**—Constructed by wrapping fine copper-gauze tightly round a piece of copper-wire, and covering the cylinder so formed with a piece of sheet-copper. The cylinders should be 3 inches long, and should fit the combustion-tube in which they are used. They should be first oxidised superficially by heating in a stream of air, and then reduced in a stream of hydrogen. If kept clean they may be used several times without being again reduced.

NITROGEN AS NITRATES AND NITRITES (Mercury Method).

This method is based upon the fact that when a concentrated solution of a nitrate is shaken up with strong sulphuric acid and mercury, the whole of the nitrogen is evolved as nitric oxide gas, according to the equation—

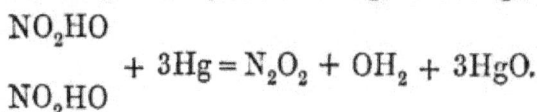

$$NO_2HO$$
$$+ 3Hg = N_2O_2 + OH_2 + 3HgO.$$
$$NO_2HO$$

The process is carried out as follows :—

The dry solid residue, which was obtained in the determination of the total solid matter in the water, is repeatedly extracted with small quantities of hot distilled water.   The extracts are filtered and the filtrate collected in a small beaker.   The solution in the beaker is then evaporated on a water-bath until its volume does not exceed 2 c.c.   This concentrated solution, which contains the whole of the nitrates, chlorides, and other soluble salts in the water, is now introduced into the tube in which the decomposition with mercury is effected. This decomposing tube (Fig. 15) is about 8 inches long, and has a bore as large as

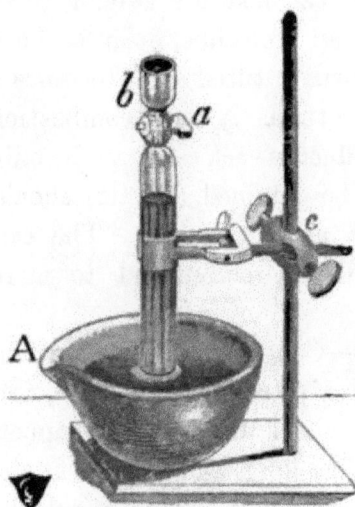

Fig. 15.

can conveniently be closed by the thumb.   It is open at one end and constricted at the other, where an accurately ground stopcock (a) is fitted ; above the latter it expands again so as to form a small cup (b).   Before use the stopcock must be greased, and the tube filled with mercury by standing the open extremity in a porcelain mercury-trough, and then, after opening the stopcock, sucking at the cup until the mercury rises to the stopcock.   The aqueous solution is introduced into the tube by pouring it into the cup, and then carefully opening the stopcock and allowing it to enter, the stopcock being again closed before any air can enter.   Should any air, however, gain access, it is readily expelled by firmly pressing the thumb into the open extremity of the tube and then momentarily opening the

stopcock. The beaker is now repeatedly rinsed with pure
strong sulphuric acid, the rinsings being similarly introduced
into the decomposing tube. A volume of sulphuric acid,
about $1\frac{1}{2}$ times as great as that of the aqueous solution,
should in all be introduced into the tube.

Any gas evolved immediately after the addition of the
sulphuric acid will consist only of carbonic anhydride, and
should be expelled. The open extremity of the tube is
now firmly closed with the thumb, and the whole tube
removed from the trough and well shaken—without, however, allowing the acid liquid to get too low in the tube.
The evolution of nitric oxide, if any nitrates or nitrites are
present, soon commences, and the tube must from time
to time be replaced in the trough to relieve the pressure
exerted upon the thumb. The shaking is continued until
no more gas is evolved, which will usually be in from three
to six minutes.

The reaction being complete, the tube is transferred to
the mercury-trough of the gas-apparatus already described,
and its contents there discharged into the laboratory
vessel, after which the volume of gas is measured in the
usual way. If $\frac{1}{2}$ litre of water has been employed for
the determination, then the volume of nitric oxide found
expresses the volume of nitrogen in 1 litre of the water,
since nitric oxide occupies twice the volume of the nitrogen it contains. From the volume found, the weight of
nitrogen per 100,000 can then be at once determined by
means of the table on p. 312.

The following example will make the method of calculation intelligible :—

## Measurement of Nitric Oxide obtained from ½ Litre of Water.

|  | Division on Measuring Tube. | Temperature. | Barometer. |
|---|---|---|---|
| Measuring tube . . | 222·5 m.m.  20 = 19·022 c.c. | 19·0 C. | 753·0 m.m. |
| Pressure tube . . | 183·8 ,,    ... | ... | 54·1 ,, |

Difference . . . 38·7 ,,    Pressure on dry gas at
                                    19·0 C. . . . . . 698·9 ,,

Plus tension of
  aqueous vapour . 16·3 ,,
Correction for capil-
  larity . . . . -·9 ,,

                   54·1 ,, to be deducted from height of barometer.

Log. of volume (19·022 c.c.) . . 1·27925

Log. of $\dfrac{0·0012562}{(1+0·00367t)760}$ for 19° C. . $\overline{6}$·18897

Log. of pressure . . . . . 2·84441

Log. of weight of gas calculated as N . $\overline{2}$·31263

Natural number . . . . . = ·02054 N per litre.
Nitrogen as nitrates and nitrites . . = 2·054 parts per 100,000.

The following is the composition of the water from a brewery-well sunk into the chalk below the London clay, and recently analysed in my laboratory :—

## Results of Analysis expressed in Parts per 100,000.

| | |
|---|---|
| Total solid matters . . . . | 78·16 |
| Organic carbon . . . . | ·230 |
| ,,    nitrogen . . . . | ·042 |
| Ammonia . . . . . | ·032 |
| Nitrogen as nitrates and nitrites . | 0 |
| Total combined nitrogen . . . | ·062 |

|  |  |  |  |  |  |  |
|---|---|---|---|---|---|---|
| Chlorine | . | . | . | . | . | 15·7 |
| Hardness { Temporary | . | . | . |  |  | 4·3 |
| Permanent | . | . | . |  |  | 3·1 |
| Total | . | . | . |  |  | 7·4 |
| Iron and alumina | . | . | . | . |  | ·160 |
| Lime | . | . | . | . | . | 2·610 |
| Magnesia | . | . | . | . | . | 1·463 |
| Silica | . | . | . | . | . | 1·310 |
| Soda | . | . | . | . | . | 34·706 |
| Potash | . | . | . | . | . | 1·588 |
| Sulphuric anhydride | . | . | . | . | . | 14·025 |
| Phosphoric　　,, | . | . | . | . |  | 0 |
| Alkalinity (equivalent to) $Na_2CO_3$ |  |  |  | . |  | 30·100 |

## DETERMINATION OF THE GASES DISSOLVED IN WATER.

It is sometimes of importance to ascertain both the quantity and the nature of the gases dissolved in water. This can most conveniently be effected by an apparatus devised by Reichardt (Fresenius, *Zeitschrift f. Analyt. Chem.* XI. p. 271). This consists of a flask *A*, of known capacity (about 1000 c.c.), which should be completely filled with the water under examination. The neck of this flask is fitted with a perforated india-rubber stopper, through which a glass tube passes connecting it with a wide-mouthed bottle *B*, of rather less capacity than the flask. This bottle *B*, which is partially filled with recently-boiled distilled water, is also fitted with an india-rubber stopper perforated with three holes, through which pass glass tubes connecting it with the flask *A* on the one hand, with another flask *C*, of about the same capacity, on the other hand, whilst through the third hole passes a delivery-tube, provided with a screw-clamp, leading to a mercury-trough *D*. The flask *C* is also partially filled with recently-boiled distilled water, and the tube connecting it with *B* is furnished with a screw-clamp. The delivery-tube only just passes through the stopper of *B*, whilst the tube from *A* passes to about one-third the height of *B* from the bottom, and the tube from *C* passes nearly to the bottom of *B*. The flask *C* is fitted with a cork perforated by tubes arranged like those of a common washbottle. By blowing into the shorter of these tubes in *C*, the bottle *B* and the tubes connected with it are completely filled with water. The clamp on the delivery-tube is now closed, whilst that on the tube between *B* and *C* is left open. The water in *A* is then carefully boiled, and the gas evolved collects in *B*, the water displaced being forced into *C*. The ebullition is continued for one hour, in which time the whole of the dissolved gases will have been expelled. The clamp on the delivery-tube is now opened, and by cautiously blowing into the flask *C*, the water is expelled from the delivery-tube, which is then placed with its extremity under the graduated receiver in which the gas is to be collected over mercury. On then continuing to blow into *C*, the gas which has collected in *B* is completely displaced and transferred through the delivery-tube into the measuring tube in the mercury-trough. On withdrawing the lamp from under *A*, the latter should completely fill with water; if it does not, the ebullition must be continued and the remaining gas collected as before.

The exact volume of the gas evolved is ascertained by accurately reading the graduation at which the mercury stands in the tube, which must be carefully calibrated. The barometric pressure and temperature of the air are also noted, and the difference in level between the mercury in the trough and in the tube measured; this difference, together with the tension of aqueous vapour for the temperature of the air, must be subtracted from the height of the barometer in order to obtain the pressure under which the dry gas occupies the volume found; the volume of the gas at 760 m.m. pressure and 0° C. can then be calulated from the formula—

$$V_0 = V_t \cdot \frac{p}{(1 + 0\cdot00367t)\,760,}$$

in which $V_0$ = volume at 0° C. and 760 m.m.

$V_t$ = volume at t° C.

$p$ = pressure on dry gas.

The absolute volume of the gas evolved being thus determined, its percentage composition can then be ascertained by transferring about 1·0 c.c. to the laboratory vessel of the gas-apparatus, and then proceeding as in the analysis of the gases obtained in the combustion of water-residues. The carbonic anhydride is first removed with caustic potash, and then the oxygen with pyrogallic acid. The residual gas is nitrogen.

The following were the quantities and composition of the dissolved gases obtained from waters of various sources by Frankland and M'Leod:—

| | | Rain Water. 15°·5 C. | Thames Water. (Grand Junction Company.) 14°·8 C. | Water from Well in Chalk 367 feet deep. 12°·1 C. |
|---|---|---|---|---|
| Volumes of gases in 100 volumes of water | Nitrogen | 1·382 | 1·397 | 2·030 |
| | Oxygen | ·673 | ·620 | ·029 |
| | Carbonic Anhydride | ·135 | 4·239 | 5·765 |
| | Total | 2·190 | 6·256 | 7·824 |

Y

# APPENDIX.

## SYMBOLS AND ATOMIC WEIGHTS OF THE ELEMENTS.

| Element. | Symbol. | Atomic Weight. | Element. | Symbol. | Atomic Weight. |
|---|---|---|---|---|---|
| Aluminium | Al | 27·02 | Mercury | Hg | 200·00 |
| Antimony | Sb | 120·0 | Molybdenum | Mo | 96·00 |
| Arsenic | As | 75·15 | Nickel | Ni | 58·74 |
| Barium | Ba | 136·84 | Niobium | Nb | 94·00 |
| Bismuth | Bi | 210·0 | Nitrogen | N | 14·01 |
| Boron | B | 11·04 | Osmium | Os | 199·03 |
| Bromine | Br | 79·76 | Oxygen | O | 15·96 |
| Cadmium | Cd | 112·04 | Palladium | Pd | 106·57 |
| Caesium | Cs | 133·00 | Phosphorus | P | 30·96 |
| Calcium | Ca | 39·90 | Platinum | Pt | 194·46 |
| Carbon | C | 11·97 | Potassium | K | 39·11 |
| Cerium | Ce | 138·24 | Rhodium | Rh | 104·21 |
| Chlorine | Cl | 35·46 | Rubidium | Rb | 85·40 |
| Chromium | Cr | 52·08 | Ruthenium | Ru | 104·40 |
| Cobalt | Co | 58·74 | Selenium | Se | 79·46 |
| Copper | Cu | 63·12 | Silver | Ag | 107·67 |
| Didymium | D | 142·44 | Silicon | Si | 28·10 |
| Erbium | E | 168·9 | Sodium | Na | 22·99 |
| Fluorine | F | 18·96 | Strontium | Sr | 87·54 |
| Gallium | Ga | 69·8 | Sulphur | S | 31·996 |
| Glucinum | Gl | 9·30 | Tantalum | Ta | 182·30 |
| Gold | Au | 196·71 | Tellurium | Te | 128·06 |
| Hydrogen | H | 1·0 | Thallium | Tl | 203·66 |
| Indium | In | 113·4 | Thorium | Th | 231·44 |
| Iodine | I | 126·54 | Tin | Sn | 118·10 |
| Iridium | Ir | 196·87 | Titanium | Ti | 50·00 |
| Iron | Fe | 56·00 | Tungsten | W | 184·00 |
| Lanthanum | La | 139·33 | Uranium | U | 240·00 |
| Lead | Pb | 206·40 | Vanadium | V | 51·35 |
| Lithium | Li | 7·00 | Yttrium | Y | 92·55 |
| Magnesium | Mg | 23·94 | Zinc | Zn | 65·16 |
| Manganese | Mn | 54·04 | Zirconium | Zr | 89·60 |

## Percentage of Pure Sulphuric Acid ($H_2SO_4$) and Sulphuric Anhydride ($SO_3$) in Aqueous Sulphuric Acid of various Specific Gravities (at 15° C.)—Bineau.

| Specific Gravity. | $H_2SO_4$ % | $SO_3$ % | Specific Gravity. | $H_2SO_4$ % | $SO_3$ % | Specific Gravity. | $H_2SO_4$ % | $SO_3$ % |
|---|---|---|---|---|---|---|---|---|
| 1·8426 | 100 | 81·63 | 1·578 | 66 | 53·87 | 1·2476 | 33 | 26·94 |
| 1·842 | 99 | 80·81 | 1·557 | 65 | 53·05 | 1·239 | 32 | 26·12 |
| 1·8406 | 98 | 80·00 | 1·545 | 64 | 52·24 | 1·231 | 31 | 25·30 |
| 1·840 | 97 | 79·18 | 1·534 | 63 | 51·42 | 1·223 | 30 | 24·49 |
| 1·8384 | 96 | 78·36 | 1·523 | 62 | 50·61 | 1·215 | 29 | 23·67 |
| 1·8376 | 95 | 77·55 | 1·512 | 61 | 49·70 | 1·2066 | 28 | 22·85 |
| 1·8356 | 94 | 76·73 | 1·501 | 60 | 48·98 | 1·198 | 27 | 22·03 |
| 1·834 | 93 | 75·91 | 1·490 | 59 | 48·16 | 1·190 | 26 | 21·22 |
| 1·831 | 92 | 75·10 | 1·480 | 58 | 47·34 | 1·182 | 25 | 20·40 |
| 1·827 | 91 | 74·28 | 1·469 | 57 | 46·53 | 1·174 | 24 | 19·58 |
| 1·822 | 90 | 73·47 | 1·4586 | 56 | 45·71 | 1·167 | 23 | 18·77 |
| 1·816 | 89 | 72·65 | 1·448 | 55 | 44·89 | 1·159 | 22 | 17·95 |
| 1·809 | 88 | 71·83 | 1·438 | 54 | 44·07 | 1·1516 | 21 | 17·14 |
| 1·802 | 87 | 71·02 | 1·428 | 53 | 43·26 | 1·144 | 20 | 16·32 |
| 1·794 | 86 | 70·10 | 1·418 | 52 | 42·45 | 1·136 | 19 | 15·51 |
| 1·786 | 85 | 69·38 | 1·408 | 51 | 41·63 | 1·129 | 18 | 14·69 |
| 1·777 | 84 | 68·57 | 1·398 | 50 | 40·81 | 1·121 | 17 | 13·87 |
| 1·767 | 83 | 67·75 | 1·3886 | 49 | 40·00 | 1·1136 | 16 | 13·06 |
| 1·756 | 82 | 66·94 | 1·379 | 48 | 39·18 | 1·106 | 15 | 12·24 |
| 1·745 | 81 | 66·12 | 1·370 | 47 | 38·36 | 1·098 | 14 | 11·42 |
| 1·734 | 80 | 65·30 | 1·361 | 46 | 37·55 | 1·091 | 13 | 10·61 |
| 1·722 | 79 | 64·48 | 1·351 | 45 | 36·73 | 1·083 | 12 | 9·79 |
| 1·710 | 78 | 63·67 | 1·342 | 44 | 35·82 | 1·0756 | 11 | 8·98 |
| 1·698 | 77 | 62·85 | 1·333 | 43 | 35·10 | 1·068 | 10 | 8·16 |
| 1·686 | 76 | 62·04 | 1·324 | 42 | 34·28 | 1·061 | 9 | 7·34 |
| 1·675 | 75 | 61·22 | 1·315 | 41 | 33·47 | 1·0536 | 8 | 6·53 |
| 1·663 | 74 | 60·40 | 1·306 | 40 | 32·65 | 1·0464 | 7 | 5·71 |
| 1·651 | 73 | 59·59 | 1·2976 | 39 | 31·83 | 1·039 | 6 | 4·89 |
| 1·639 | 72 | 58·77 | 1·289 | 38 | 31·02 | 1·032 | 5 | 4·08 |
| 1·637 | 71 | 57·95 | 1·281 | 37 | 30·20 | 1·0256 | 4 | 3·26 |
| 1·615 | 70 | 57·14 | 1·272 | 36 | 29·38 | 1·019 | 3 | 2·445 |
| 1·604 | 69 | 56·32 | 1·264 | 35 | 28·57 | 1·013 | 2 | 1·63 |
| 1·592 | 68 | 55·59 | 1·256 | 34 | 27·75 | 1·0064 | 1 | 0·816 |
| 1·580 | 67 | 54·69 | ... | ... | ... | ... | ... | ... |

PERCENTAGE OF PURE HYDROCHLORIC ACID (HCl) IN AQUEOUS HYDROCHLORIC ACID OF VARIOUS SPECIFIC GRAVITIES (AT 15° C.)—Ure.

| Specific Gravity. | HCl %. | Specific Gravity. | HCl %. | Specific Gravity. | HCl %. | Specific Gravity. | HCl %. |
|---|---|---|---|---|---|---|---|
| 1·2000 | 40·777 | 1·1515 | 30·582 | 1·1000 | 20·388 | 1·0497 | 10·194 |
| 1·1982 | 40·369 | 1·1494 | 30·174 | 1·0980 | 19·980 | 1·0477 | 9·786 |
| 1·1964 | 39·961 | 1·1473 | 29·767 | 1·0960 | 19·572 | 1·0457 | 9·379 |
| 1·1946 | 39·554 | 1·1452 | 29·359 | 1·0939 | 19·165 | 1·0437 | 8·971 |
| 1·1928 | 39·146 | 1·1431 | 28·951 | 1·0919 | 18·757 | 1·0417 | 8·563 |
| 1·1910 | 38·738 | 1·1410 | 28·544 | 1·0899 | 18·349 | 1·0397 | 8·155 |
| 1·1893 | 38·330 | 1·1389 | 28·136 | 1·0879 | 17·941 | 1·0377 | 7·747 |
| 1·1875 | 37·923 | 1·1369 | 27·728 | 1·0859 | 17·534 | 1·0357 | 7·340 |
| 1·1857 | 37·516 | 1·1349 | 27·321 | 1·0838 | 17·126 | 1·0337 | 6·932 |
| 1·1846 | 37·108 | 1·1328 | 26·913 | 1·0818 | 16·718 | 1·0318 | 6·524 |
| 1·1822 | 36·700 | 1·1308 | 26·505 | 1·0798 | 16·310 | 1·0298 | 6·116 |
| 1·1802 | 36·292 | 1·1287 | 26·098 | 1·0778 | 15·902 | 1·0279 | 5·709 |
| 1·1782 | 35·884 | 1·1267 | 25·690 | 1·0758 | 15·494 | 1·0259 | 5·301 |
| 1·1762 | 35·476 | 1·1247 | 25·282 | 1·0738 | 15·087 | 1·0239 | 4·893 |
| 1·1741 | 35·068 | 1·1226 | 24·874 | 1·0718 | 14·679 | 1·0220 | 4·486 |
| 1·1721 | 34·660 | 1·1206 | 24·466 | 1·0697 | 14·271 | 1·0200 | 4·078 |
| 1·1701 | 34·252 | 1·1185 | 24·058 | 1·0677 | 13·863 | 1·0180 | 3·670 |
| 1·1681 | 33·845 | 1·1164 | 23·650 | 1·0657 | 13·456 | 1·0160 | 3·262 |
| 1·1661 | 33·437 | 1·1143 | 23·242 | 1·0637 | 13·049 | 1·0140 | 2·854 |
| 1·1641 | 33·029 | 1·1123 | 22·834 | 1·0617 | 12·641 | 1·0120 | 2·447 |
| 1·1620 | 32·621 | 1·1102 | 22·426 | 1·0597 | 12·233 | 1·0100 | 2·039 |
| 1·1599 | 32·213 | 1·1082 | 22·019 | 1·0577 | 11·825 | 1·0080 | 1·631 |
| 1·1578 | 31·805 | 1·1061 | 21·611 | 1·0557 | 11·418 | 1·0060 | 1·124 |
| 1·1557 | 31·398 | 1·1041 | 21·203 | 1·0537 | 11·010 | 1·0040 | 0·816 |
| 1·1536 | 30·990 | 1·1020 | 20·796 | 1·0517 | 10·602 | 1·0020 | 0·408 |

PERCENTAGE AMOUNT OF NITRIC ACID (HNO$_3$) CONTAINED IN AQUEOUS SOLUTIONS OF VARIOUS SPECIFIC GRAVITIES (at 15° C.)—(Kolb, *Ann. Ch. Phys.* [4] 136).

| Specific Gravity. | HNO$_3$ % | Specific Gravity. | HNO$_3$ % | Specific Gravity. | HNO$_3$ % | Specific Gravity. | HNO$_3$ % |
|---|---|---|---|---|---|---|---|
| 1·530 | 100·00 | 1·460 | 80·00 | 1·368 | 58·88 | 1·237 | 37·95 |
| 1·530 | 99·84 | 1·456 | 79·00 | 1·363 | 58·00 | 1·225 | 36·00 |
| 1·530 | 99·72 | 1·451 | 77·66 | 1·358 | 57·00 | 1·218 | 35·00 |
| 1·529 | 99·52 | 1·445 | 76·00 | 1·353 | 56·10 | 1·211 | 33·86 |
| 1·523 | 97·89 | 1·442 | 75·00 | 1·346 | 55·00 | 1·198 | 32·00 |
| 1·520 | 97·00 | 1·438 | 74·01 | 1·341 | 54·00 | 1·192 | 31·00 |
| 1·516 | 96·00 | 1·435 | 73·00 | 1·339 | 53·81 | 1·185 | 30·00 |
| 1·514 | 95·27 | 1·432 | 72·39 | 1·335 | 53·00 | 1·179 | 29·00 |
| 1·509 | 94·00 | 1·429 | 71·24 | 1·331 | 52·33 | 1·172 | 28·00 |
| 1·506 | 93·01 | 1·423 | 69·96 | 1·323 | 50·99 | 1·166 | 27·00 |
| 1·503 | 92·00 | 1·419 | 69·20 | 1·317 | 49·97 | 1·157 | 25·71 |
| 1·496 | 91·00 | 1·414 | 68·00 | 1·312 | 49·00 | 1·138 | 23·00 |
| 1·495 | 90·00 | 1·410 | 67·00 | 1·304 | 48·00 | 1·120 | 20·00 |
| 1·494 | 89·56 | 1·405 | 66·00 | 1·298 | 47·18 | 1·105 | 17·47 |
| 1·488 | 88·00 | 1·400 | 65·07 | 1·295 | 46·64 | 1·089 | 15·00 |
| 1·486 | 87·45 | 1·395 | 64·00 | 1·284 | 45·00 | 1·077 | 13·00 |
| 1·482 | 86·17 | 1·393 | 63·59 | 1·274 | 43·53 | 1·067 | 11·41 |
| 1·478 | 85·00 | 1·386 | 62·00 | 1·264 | 42·00 | 1·045 | 7·22 |
| 1·474 | 84·00 | 1·381 | 61·21 | 1·257 | 41·00 | 1·022 | 4·00 |
| 1·470 | 83·00 | 1·374 | 60·00 | 1·251 | 40·00 | 1·010 | 2·00 |
| 1·467 | 82·00 | 1·372 | 59·59 | 1·244 | 39·00 | 0·999 | 0·00 |
| 1·463 | 80·96 | ... | ... | ... | ... | ... | ... |

PERCENTAGE AMOUNT OF CAUSTIC POTASH ($K_2O$) IN AQUEOUS SOLUTIONS OF VARIOUS SPECIFIC GRAVITIES (AT 15° C.)—Tünnermann, *N. Tr.* xviii. 2, 5.

| Specific Gravity. | $K_2O$ %. | Specific Gravity. | $K_2O$ %. |
|---|---|---|---|
| 1·3300 | 28·290 | 1·1437 | 14·145 |
| 1·3131 | 27·158 | 1·1308 | 13·013 |
| 1·2966 | 26·027 | 1·1182 | 11·882 |
| 1·2805 | 24·895 | 1·1059 | 10·750 |
| 1·2648 | 23·764 | 1·0938 | 9·619 |
| 1·2493 | 22·632 | 1·0819 | 8·487 |
| 1·2342 | 21·500 | 1·0703 | 7·355 |
| 1·2268 | 20·935 | 1·0589 | 6·224 |
| 1·2122 | 19·803 | 1·0478 | 5·002 |
| 1·1979 | 18·671 | 1·0369 | 3·961 |
| 1·1839 | 17·540 | 1·0260 | 2·829 |
| 1·1702 | 16·408 | 1·0153 | 1·697 |
| 1·1568 | 15·277 | 1·0050 | 0·5658 |

PERCENTAGE AMOUNT OF CAUSTIC SODA (Na$_2$O) IN
AQUEOUS SOLUTIONS OF VARIOUS SPECIFIC GRAVI-
TIES (AT 15° C.)—Tünnermann.

| Specific Gravity. | Na$_2$O %. | Specific Gravity. | Na$_2$O %. | Specific Gravity. | Na$_2$O %. |
|---|---|---|---|---|---|
| 1·4285 | 30·220 | 1·2912 | 19·945 | 1·1428 | 9·670 |
| 1·4193 | 29·616 | 1·2843 | 19·341 | 1·1330 | 9·066 |
| 1·4101 | 29·011 | 1·2775 | 18·730 | 1·1233 | 8·462 |
| 1·4011 | 28·407 | 1·2708 | 18·132 | 1·1137 | 7·857 |
| 1·3923 | 27·802 | 1·2642 | 17·528 | 1·1042 | 7·253 |
| 1·3836 | 27·200 | 1·2578 | 16·923 | 1·0948 | 6·648 |
| 1·3751 | 26·594 | 1·2515 | 16·379 | 1·0855 | 6·044 |
| 1·3668 | 25·989 | 1·2453 | 15·714 | 1·0764 | 5·440 |
| 1·3586 | 25·385 | 1·2392 | 15·110 | 1·0675 | 4·835 |
| 1·3505 | 24·780 | 1·2280 | 14·500 | 1·0587 | 4·231 |
| 1·3426 | 24·176 | 1·2178 | 13·901 | 1·0500 | 3·626 |
| 1·3349 | 23·572 | 1·2058 | 13·297 | 1·0414 | 3·022 |
| 1·3273 | 22·967 | 1·1948 | 12·692 | 1·0330 | 2·418 |
| 1·3198 | 22·363 | 1·1841 | 12·088 | 1·0246 | 1·813 |
| 1·3143 | 21·894 | 1·1734 | 11·484 | 1·0163 | 1·209 |
| 1·3125 | 21·758 | 1·1630 | 10·879 | 1·0081 | 0·604 |
| 1·3053 | 21·154 | 1·1528 | 10·275 | 1·0040 | 0·302 |
| 1·2982 | 20·550 | ... | ... | ... | ... |

## APPENDIX.

Percentage Amount of Ammonia ($NH_3$) in Aqueous Solutions of the Gas of various Specific Gravities (at 14° C.)—Carius.

| Specific Gravity. | $NH_3$ %. | Specific Gravity. | $NH_3$ %. |
| --- | --- | --- | --- |
| 0·8844 | 36 | 0·9314 | · 18 |
| 0·8864 | 35 | 0·9347 | 17 |
| 0·8885 | 34 | 0·9380 | 16 |
| 0·8907 | 33 | 0·9414 | 15 |
| 0·8929 | 32 | 0·9449 | 14 |
| 0·8953 | 31 | 0·9484 | 13 |
| 0·8976 | 30 | 0·9520 | 12 |
| 0·9001 | 29 | 0·9556 | 11 |
| 0·9026 | 28 | 0·9593 | 10 |
| 0·9052 | 27 | 0·9631 | 9 |
| 0·9078 | 26 | 0·9670 | 8 |
| 0·9106 | 25 | 0·9709 | 7 |
| 0·9133 | 24 | 0·9749 | 6 |
| 0·9162 | 23 | 0·9790 | 5 |
| 0·9191 | 22 | 0·9831 | 4 |
| 0·9221 | 21 | 0·9873 | 3 |
| 0·9251 | 20 | 0·9915 | 2 |
| 0·9283 | 19 | 0·9959 | 1 |

THE END.